HAPPY SAILING
GENE & NANCY

# Alaska Blues

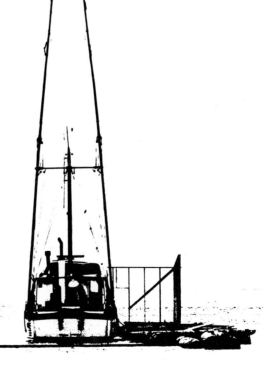

# Alaska Blues
## A Fisherman's Journal

## Joe Upton

ALASKA NORTHWEST PUBLISHING COMPANY
Anchorage, Alaska

Second printing 1979

Library of Congress cataloging in publication data:
Upton, Joe, 1946-
    Alaska blues.
    1.   Pacific salmon fisheries—Alaska.    I.   Title.
SH336.5.S24U68        639'.27'55       77-23925
ISBN 0-88240-098-3

Design by Dianne Hofbeck
CartoGraphics by Jon. Hersh

Alaska Northwest Publishing Company
Box 4-EEE, Anchorage, Alaska 99509
Printed in U.S.A.

*For all my friends in the fishery,*
*without whom this would have been impossible*

*For Laura, who taught me layouts*
*and gave me so much of her time*

*But most of all for Susanna,*
*who did it all with me*

*"I keep telling myself that some year I'll just bag it, and spend the whole season . . . in a little boat, maybe a sloop or something, poking around. But then each spring comes and I have to go fishing."*

# Contents

*"All up and down the coast . . .
the salmon fleet lies at the
docks . . . covered with snow,
closed and silent, laid up
for the winter."*

# The Beginning

In December and January the winter lies hard on the land in Southeastern Alaska. The bays are silent and the channels and passages empty. Up in Icy Strait and out on Baranof and Chichagof islands, the inner bays freeze over solid, sometimes from December until March. The canneries and the fish camps are closed down and deserted except for caretakers making their solitary rounds. Many men from the area take their families and go south, and in some of the smaller communities the only sign of life is a light in a window or a plume of wood smoke from a chimney. All up and down the coast, from Washington to Alaska, the salmon fleet lies at the docks; boats are tied two and three deep, covered with snow, closed and silent, laid up for the winter. These are quiet months in the North Country.

But then comes March with the promise of spring. The bays thaw, the snow turns to rain, and the docks and harbors start to come to life as the

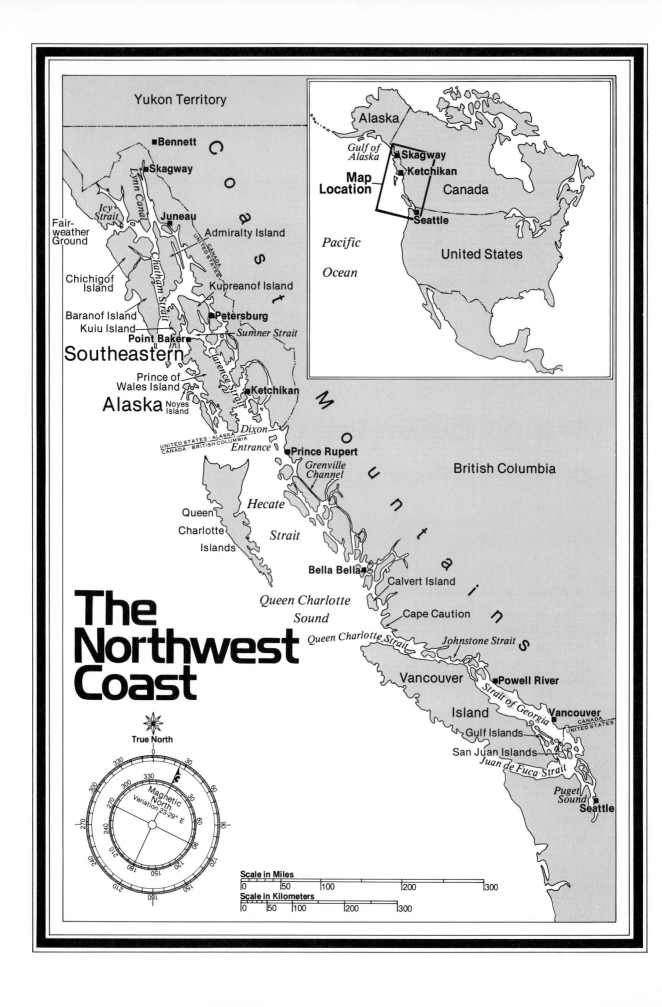

Yukon Territory

■ **Bennett**

■ **Skagway**

*Lynn Canal*

*Icy Strait*

■ **Juneau**

Fairweather Ground

Admiralty Island

*Chatham Strait*

Chichigof Island

Kupreanof Island

Baranof Island
Kuiu Island

■ **Petersburg**

**Point Baker** ■

*Sumner Strait*

## Southeastern

Prince of Wales Island

*Clarence Strait*

■ **Ketchikan**

## Alaska

Noyes Island

*Dixon*

UNITED STATES · ALASKA
CANADA · BRITISH COLUMBIA

*Entrance*

■ **Prince Rupert**

*Grenville Channel*

British Columbia

*Hecate*

Queen Charlotte Islands

*Strait*

**Bella Bella** ■

Calvert Island

*Queen Charlotte Sound*

Cape Caution

# The Northwest Coast

*Queen Charlotte Strait*

*Johnstone Strait*

Vancouver

**Powell River** ■

Island

*Strait of Georgia*

**Vancouver** ■

CANADA
UNITED STATES

Gulf Islands

San Juan Islands

*Juan de Fuca Strait*

*Puget Sound*

**Seattle** ■

True North

Magnetic North
Variation 23-29° E

**Scale in Miles**
0    50    100         200              300

**Scale in Kilometers**
0   50  100      200        300

### Inset map

Alaska

*Gulf of Alaska*

■ **Skagway**

■ **Ketchikan**

**Map Location**

Canada

■ **Seattle**

*Pacific*

*Ocean*

United States

*The packer* Orient *lies at Baker Inlet, Grenville Channel.*

men begin to fit out their boats for the season ahead. Soon from the harbors up north and from all the Puget Sound ports in Washington, the Alaska fleet starts to leave. The first to go are a handful of packers and a few seiners, husky boats bound for the early herring runs. (Packers—or tenders—are boats that buy the catch on the fishing grounds and take it to a cannery.) But only a few boats leave so early, for March can still be winter up north, and the bulk of the fleet is waiting for the salmon to start running in from the ocean.

In April, the salmon fleet begins to leave; first a trickle but soon a flood of boats is heading north. It starts with the big outside trollers—boats sturdy enough to leave the protection of the Inside Passage and work in the open ocean. Forty and fifty feet long, with high bows and sweeping lines, the outside trollers are built for the rugged ocean fishery at Noyes Island and the Fairweather Ground, 40 miles off the Alaskan coast. The season opens April 15 on the North Coast, and many boats travel a thousand miles to make it. Sometimes they arrive only to lie in the harbors, sailing back and forth on their anchors in the gusts as the gales march by on the outside. When spring is late in coming, even the largest lie in and wait.

Into May the trolling fleet continues to stream north, the smallest boats last, some of them hardly more than skiffs. Some boats are bound for the outside—the open sea—to fish the coast when the weather's fair, but many more are headed for the inside, to the hundreds of bays and holes and nooks of the Inside Passage where the smallest boat can hide and still fish when it blows.

By the middle of the month there is almost a procession from Puget Sound, as the rest of the salmon fleet begins to move out. Each tide sees a

few more boats headed north, decks often piled high with gear and boxes yet to be stowed.

But June is the busiest month, when the seine and gill-net fleets leave. Their season starts around the end of the month, and hundreds of boats travel north every year for it. They leave four and five at once, newly painted, and it's a festive time, with wives and families crowding the docks to see them off.

For a few weeks in June it seems as if all the channels north, through Canada and into Alaska, are full of boats, running in pairs and in groups, taking advantage of the good weather and long hours of daylight to run 18 and 20 hours a day. In the evenings the boats trickle into the anchorages until deep in the night; many run straight through, day and night. In June you might anchor early and alone in some cove in Canada, a hundred miles from the nearest town or road, and wake up before dawn in surprise at the sight of five or ten other boats in the anchorage with you, all riding silently at their anchors.

By the end of the month the exodus is pretty well over, the Alaska fleet gone, and for a few months many of the harbors in Puget Sound seem almost empty.

Up north the fleet spreads out, to bays and coves and channels almost without number. Bays that lie silent for 11 months of the year become, for a few brief weeks, bright cities of boats fishing day and night while the salmon run. Then the whole coast is alive as the boats and men try to make a year's living in a season.

The canneries start up, and towns that have slept through the winter now begin to boom and fill up with seasonal workers and fishermen. In a big season the money's easy, the canneries work double shifts, and the bars are open day and night.

But summer is brief up north, and fall comes early. By the middle of September a steady stream of boats is winding its way south again. All through October the boats return, and into November if the weather's been bad. They run down singly and in groups, splitting up when they get to Puget Sound, to go to dozens of harbors and sloughs. When they return, often one by one and late at night, it's with little fanfare. In the morning you walk the docks and a few more are back, rust-streaked, their paint faded and chipped. The boats are already shut up and deserted, the crews paid off and gone home, scattered for the winter. By the middle of November, all the boats are back.

*Right—Company boats laid up for the winter at Craig, Alaska. Opposite—A packer early in the season in the Inside Passage.*

After a long season, six or seven months on the boat, and maybe a dirty trip home, you're ready to quit. You want to just tie the boat up and do something else for a while. Maybe go hunting, take a trip somewhere or just spend time at home with your family or your friends.

But then the dead of winter passes, you get a couple of warm, clear days, and all of a sudden you find yourself wandering down to the boat, seeing what's to be done, and thinking about going north again.

That lonely country gets in your blood after a while. You go up there for a couple of seasons and pretty soon, when the warm weather comes, that's where you want to be. For many of the men in the fishery, for 20 or 30 years—ever since they were teenagers and maybe before—spring has always meant getting ready to go north. In different boats, to different places, but always to Alaska.

Some quit for a while, try other jobs, try staying south, being with their families. But for many it doesn't work. Spring comes and they think about being up north, lying in some lonely bay on a gray morning hanging on the net, or jogging the boat into a squall with nothing around but wind and sky and water. When the time comes, they say their good-bys and head north.

This book is the journal of a salmon season in Alaska and all that that means—the trip north to that magnificent place, God's country, the fishing in steep, tree-lined fjords, the glorious days on the beach, and the stormy trip home.

*A*PRIL *12*—Today, on as fine a Seattle morning as I have seen this year, the mountains pink both to east and west, we slip through the locks and down into the salt water of Puget Sound.

Decks and quarters are still littered with gear and odds and ends to be stowed; the big hold is full to the hatch cover with nets, buoys and cases of food and oil—everything that we need for a long season ahead. We could have stayed another day and finished up all the little jobs that still had to be done. But with a fair tide and a clear sky, it was time to go. I swung my compass at the outer buoy, pulled the first chart off the roll, and pointed the bow north.

So it all begins again. After months of preparation, the weeks blurring one into another, we're off, to travel almost a thousand miles north to fish for salmon, to make a year's living in just a few months. We've done all that we can to be ready and now we're on our way, but the season ahead is far from certain.

My boat is the *Doreen*, a fiberglass workboat, 32 feet long. With a big power reel on the afterdeck and hydraulic gurdies on either side of the reel, we're rigged to gill-net and to troll. (Gurdies are special winches for hauling in trolling lines.) Amidships is a roomy insulated hold, forward is the pilothouse with a little galley, and below is the fo'c'sle with a double berth, a toilet and two well-stocked bookcases. With plenty of storage space, the accommodations are simple but comfortable for my wife Susanna, our dog Sam, and me. We're in for a long season. We have two radios: a marine radio and a citizen's band radio. Diesel-powered and radar-equipped, the boat is neither plainer nor fancier than most of the fleet that heads north every spring.

Pushed by tide, warmed by sun, we boiled out of the sound with the big ebb. The strait opened up ahead, and Susanna took the wheel while I went over the whole boat. Listened to the engine, checked the stuffing box, tuned the radar, and tried the radios and depth sounders. Better to find problems now, with time and weather in my favor, than be surprised in some bad spot with my hands full and no place to go.

Kestrel *and* Doreen, *two examples of the Alaskan gillnetter-troller, ready for a long season.*

Kestrel *follows* Doreen, *as the two roll their way across the Strait of Georgia, inside Vancouver Island.*

To the south was the snowy wall of the Olympic Mountains; to the east, the North Cascades; to the west, the Juan de Fuca Strait and the distant Pacific Ocean. The tide turned, and we caught the push of the flood tide through the San Juan Islands and deep into Canada. An hour before dusk we slipped alongside the empty float at Musgrave Landing on Saltspring Island, where we lie now on a warm spring evening with not a ripple on the water nor a breeze in the trees.

Alongside is the *Kestrel*, with Bruce and his wife, Kathy, our fishing partners from seasons past. We walked through the still woods tonight and sat on the bank above the cove watching the sun go down in the hills of Vancouver Island across the channel. Now the busy weeks of fitting out are past, and the whole season is ahead.

*APRIL 13*—Up at 5:30 a.m. to walk the dog, our breath white this chill clear morning. But I suspicion a change from the falling glass and the high clouds to the south. That dog—don't know who enjoys these trips more, us or him—all day in the bunk and a new place to smell and run each night.

We spent the morning in narrow channels; here and there we saw summer homes and cottages tucked into the coves between the high hills. But today's the last of them; tomorrow we'll be past the last road, up in that lonely country where settlements are few. Under the power lines at Dodd Narrows, with the tide pushing us on, we went through the tiny gap in the islands and out into the Strait of Georgia, where the day grew chill and the sky clouded. Behind us the Gulf Islands were lost in the haze, and ahead loomed the jagged line of the Coast Mountains. A breeze from the south raised a dirty chop, across which we quartered, rolling heavily; Susanna and Sam headed for the bunk. The summer homes and warm morning seemed far away, with the wind making up all the time. Dirty going in the rip off Thormanby Islands; we took a queer one aboard before making our turn and putting the sea on the stern for the run up inside Texada Island. To the east, Agamemnon Channel and Jervis Inlet wound away deep into the gloomy interior between the steep peaks. Powell River at 6 p.m., the last big mainland town, with the lights of the mill ablaze in the stormy dusk and the rain in spits, hissing across the water with the rising wind. The night came black, with the gusts of wind laying us over at times, but an hour after dark we passed into the sheltered waters of Desolation Sound, with places to hole up on all sides. Only the eerie outline on the radar screen showed the dark shores, awesome mountains and deep inlets that cut off the coast road from the water. We spotlighted our way alongside the half-sunk float at Redonda Bay at 11 p.m. in the wind and rain. The second day out of Seattle and already the vast, empty North Country has closed in around us.

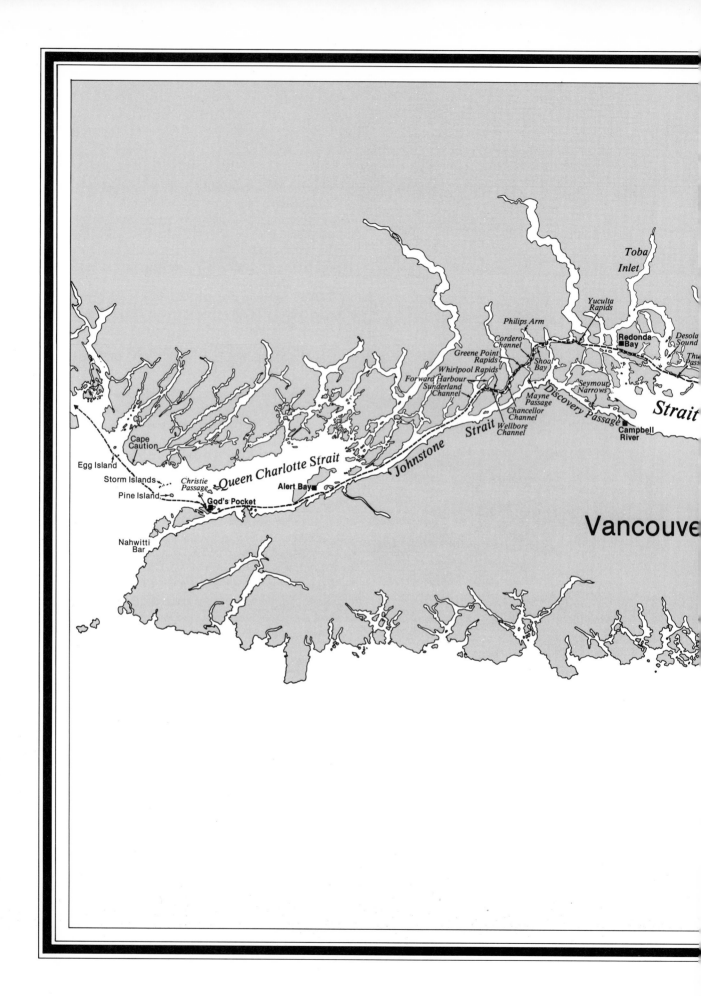

# The Inside Passage

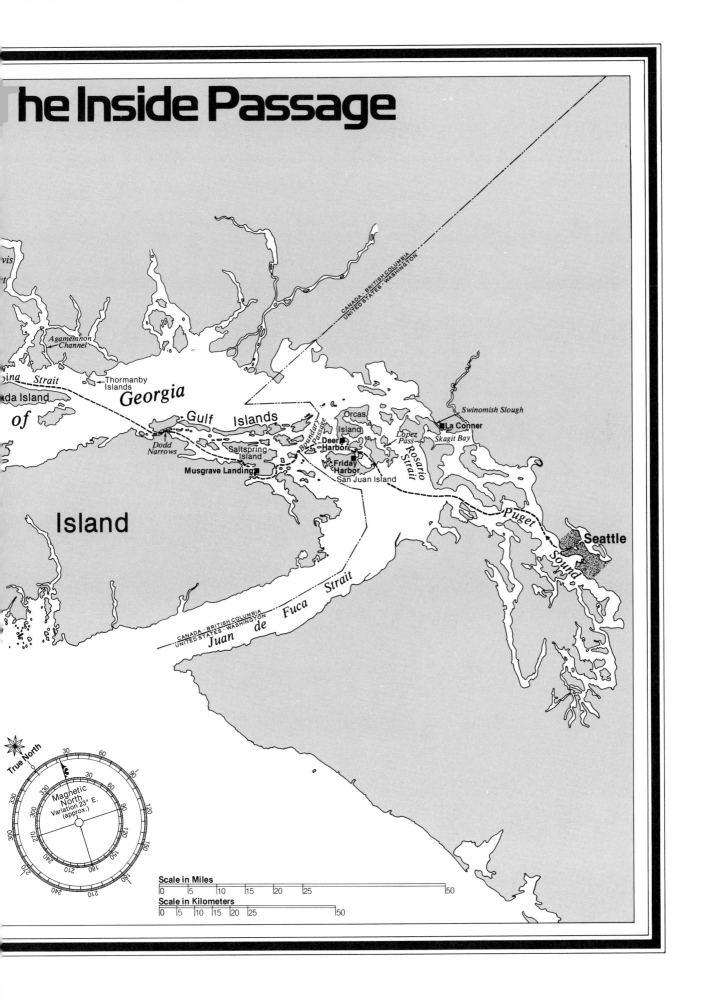

Agamemnon Channel

Thormanby Islands

ing Strait

da Island

Georgia

Gulf Islands

Dodd Narrows

Saltspring Island

Musgrave Landing

Boundary Passage

Deer Harbor

Friday Harbor

San Juan Island

Orcas Island

Lopez Pass

Swinomish Slough

La Conner

Skagit Bay

Rosario Strait

CANADA - BRITISH COLUMBIA
UNITED STATES - WASHINGTON

of

Island

Puget Sound

Seattle

Juan de Fuca Strait

CANADA - BRITISH COLUMBIA
UNITED STATES - WASHINGTON

True North

Magnetic North
Variation 23° E.
(approx.)

Scale in Miles
0   5   10   15   20   25                    50

Scale in Kilometers
0   5   10  15  20  25              50

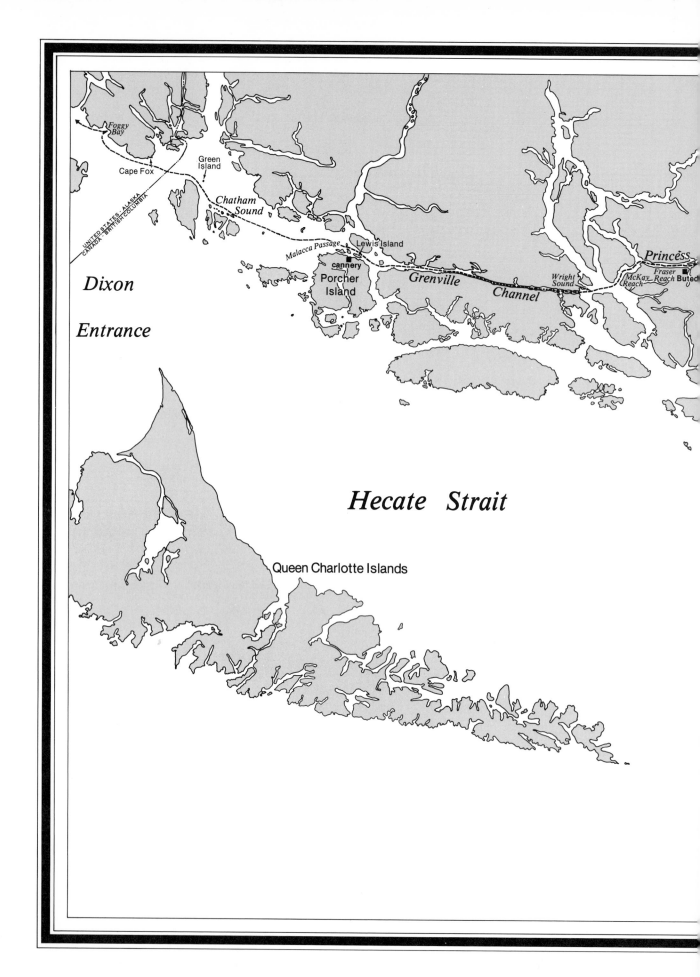

Foggy
Bay

Cape Fox

Green
Island

Chatham
Sound

UNITED STATES, ALASKA
CANADA, BRITISH COLUMBIA

*Dixon*

*Entrance*

Malacca Passage

Lewis Island

cannery

Porcher
Island

*Grenville*

*Channel*

Wright
Sound

*Princess*

Fraser
Reach

McKay
Reach

Buted

*Hecate   Strait*

Queen Charlotte Islands

# The Inside Passage

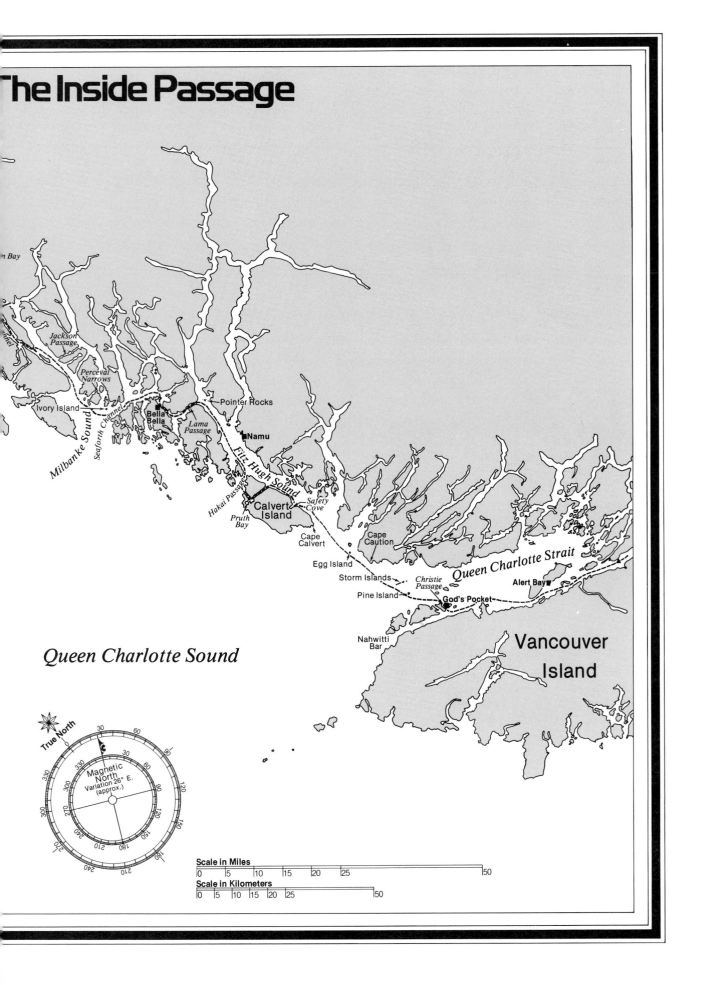

Jackson Passage

Perceval Narrows

Ivory Island

Milbanke Sound

Seaforth Channel

Bella Bella

Lama Passage

Pointer Rocks

Namu

Fitz Hugh Sound

Hakai Passage

Calvert Island

Safety Cove

Pruth Bay

Cape Calvert

Cape Caution

Egg Island

Storm Islands

Christie Passage

Pine Island

God's Pocket

Queen Charlotte Strait

Alert Bay

Nahwitti Bar

Vancouver Island

*Queen Charlotte Sound*

True North

Magnetic North
Variation 26° E.
(approx.)

30 60 90 120 150 180 210 240 270 300 330

**Scale in Miles**
0   5   10   15   20   25                    50

**Scale in Kilometers**
0   5   10   15   20   25              50

*"What gloomy canyons they
seemed today, with the clouds
pressing down . . . and the rain
beating in heavy slants
on our world of dark green hills
and lead-gray water."*

# The Inside Passage to Alaska

The way north is through the sheltered waters of the Inside Passage. North of Cape Spencer, Alaska, and south of Cape Flattery, Washington, the North Pacific beats on a lonely and forbidding coast that has few harbors. Even large vessels travel with caution.

But between the two capes lie a thousand miles of sheltered waters, with many secure harbors. The tides are large, the currents swift, many dangers unmarked, and the weather is sometimes violent, but it's inside, and seldom are you more than a few hours from a good harbor. Without the Inside Passage, the fleets of smaller boats that travel to Alaska every year to fish never would have developed. With it, in the late spring and summer, even the smallest boat can travel to Alaska in relative safety and cross open water only twice in 600 miles.

For the first part of the trip north, the land changes rapidly. For a day you run past the towns and vacation-home settlements of Puget Sound

*Totem on Chief Shakes Island, Wrangell.*

and the Canadian Gulf Islands. Then you cross the Strait of Georgia, and soon the only settlements are logging and fishing communities. The shores become steeper and the hills higher and darker, often rising right up into the overcast from the water. The weather changes, and the warm springs and bright summers of the Puget Sound country give way to the gloomy and overcast summers of coastal British Columbia. Hardly more than a hundred miles north of the Canadian border, the coast road ends, cut off from the water by high mountains and deep inlets, and from here on you're pretty much on your own, with long distances between settlements, and boat and floatplane the only means of travel. The hills are so steep and the forests so thick that even walking on the shore is impossible in places.

The traditional steamer route is along the inside shore of Vancouver Island, across open water at Queen Charlotte Sound, and then back inside. The route then plunges deep into the interior through canyonlike channels, not venturing into open water again until crossing Dixon Entrance, the Alaska border.

South of Queen Charlotte Sound, there has been extensive logging in many places, especially on Vancouver Island. But to the north of the sound are some of the wildest and most remote areas of coastal British Columbia. It is easy to travel for days with only the sight of another boat or a light on a point to show that anyone has ever been there before. On all sides, deep and gloomy inlets wind far back into the hills to harbors and bays that sometimes must see years pass without a visitor.

On a bright day, running a boat up these channels—with a fair sky overhead and hill after hill slipping away astern—can be exhilarating. But on a stormy day, with low clouds pressing down on the water and nothing but dark walls rising on all sides, it can seem as if you're traveling without end through lost and empty country.

For the small boat, not in a hurry, there are many side channels and passages that offer even more shelter than the traditional routes, especially south of Queen Charlotte Sound. Many small vessels, instead of following the Vancouver Island shore through Discovery Passage and Seymour Narrows, prefer a more easterly route. This is by way of Malaspina Strait, Thulin Passage, Desolation Sound, Yuculta Rapids, Cordero Channel, Greene Point Rapids and on to Johnstone Strait through Sunderland Channel. It is a longer route, with rushing tides and places where you must wait for slack water, and it is a more intricate route, but there are good places to stop, and it shortens what can be a long and windy buck down Johnstone Strait.

North of Queen Charlotte Sound, nearly all vessels follow the steamer course, which lies far from the ocean and is well sheltered. The channels are for the most part straight, with bold shores and few places to anchor, but here and there in hidden coves are settlements and disused canneries with rotting floats where boats can lay.

Years ago, the coast was much more populated than today. Boats were slow and refrigeration uncertain, so the canneries were located wherever there was a heavy run of fish. Many of the larger inlets had little river valleys at their heads, with farming communities that had few outlets for their products; so people there lived more by barter than by cash. But then came refrigeration, and faster boats, and the many outlying canneries were consolidated into a few in the towns where labor was cheaper. The government ceded the big inlets to the logging companies and closed them to fishing. Now all that's left are the ruins of former settlements and stands of alders and hardwoods that grew up when the evergreens were cut down. In the fall you can run for hours along shores of dark evergreens, and then suddenly, in a little cove or bay, see a stand of alders, bright yellow and orange, the start of second growth telling of some settlement long since disappeared.

It is some 600 miles from Puget Sound to Alaska. A fast seiner running day and night without any stops can make it in two and a half days. The large boats often travel this way, but the small ones, often with only one or two people on board, usually just run during the day and hole up somewhere at night. The great tides pick up logs and drift from the

*Wind-driven spray on the Inside Passage in British Columbia.*

beaches, and at night these make traveling hazardous. Running only in daylight hours, a small boat makes the trip in less than a week if the weather is good.

For some, the trip north in the spring can be a relaxing time. After weeks and perhaps months of hectic work on the boat, of seeing spring all around and not having time to enjoy it, the trip can be a real break before fishing starts. Running with friends, stopping early in the evening and going ashore, or perhaps even laying over a day and digging clams and oysters, or just poking around, is a good way to start a long season.

But more often it is the other way around. The last week ashore is just a blur of last-minute shopping and endless jobs to be done. When you finally get off, the boat's still a mess with stuff to be put away and a dozen jobs to be finished on the trip north. You're behind schedule, and the only way to get up there and make the opening of the season is to run 20 hours a day or more and hope that you don't hit anything at night.

And by the time many people head back in the fall, they just want to get home and don't really care too much about sightseeing. By then the weather has usually turned, and you have to travel when you can. After the first of September the days get rapidly shorter, and the fall gales build up offshore and sweep down the coast. Often they come back to back, one right after the other, and up and down the coast the anchorages fill up and the bottles and the cards come out, for then even the sheltered inside channels are impassable. It can be a long trip in the fall, perhaps laying up one day for every day that you run. Once a friend of ours was running in Grenville Channel when it blew up, the wind screaming down the narrow reaches. There was no place to go and the channel was too deep to anchor. He found a bight in the shore and tied his boat between two trees; the boat surged back and forth on the lines for two days.

The trip home from a season up north can mean running day after day without seeing another boat and lying in the evenings at deserted towns and canneries, perhaps having to roust out some sleepy caretaker to buy fuel. And it can mean getting blown in someplace for a couple of days somewhere along the line. After traveling for 8 or 10 days or more—and every day a fight with the weather—you come at last to two cities, Powell River and Campbell River, big mill towns. A day later you're home. Then Alaska seems very far away indeed.

I've made the trip many times, and seen the Inside Passage in most of its moods, fall and spring. But each time I travel it, north or south, I'm struck again by the lonely beauty of the coast.

*Opposite—Near the Nahwitti Channel, off the northern tip of Vancouver Island.*

*"For the small boat, not in a hurry, there are many side channels and passages that offer even more shelter than the traditional routes."* This troller is in Desolation Sound.

*APRIL 14*—A peach of a day, clear and warm, following the storm that blew away in the night. We had an early start in mind, but gave it up to walk in the woods, past the tumble-down buildings of an old cannery.

Some days entertainment for Sam is rocks; some days, sticks. It was rocks today, and I threw until my arm was sore—into the trees, into the bushes, even into the creek, and he brought back the right rock every time. Finally I threw it into the bay, and in he went and circled the spot where the rock sank, swimming and barking all at once, and that's some trick.

Spent the morning lazily, drinking coffee out on deck and talking, taking our first morning off in over a month. Above the cove, an eagle circled, and gulls swooped for herring. There were oysters on the rocks, and we picked a bucket in a few minutes, just like berries. Built a fire and steamed up a mess of clams, too. Up north sometimes we'll spend a whole summer without taking off our wool shirts. By leaving Puget Sound early in the spring and not getting back until after the fall rains have started, we miss a lot, and today brought that all back. So we savored the warm day in that wild spot. But we were off at 6 p.m. when the tide served, through Yuculta Rapids and Devils Hole, evil places with oily, sinister whirlpools even at slack water. Two hours' run through twisting narrow channels between high hills, and hardly a sign of man. Reached Shoal Bay and the old government float in the pink dusk. At the head of the bay a defunct lodge sprawls across the grassy flats; behind, the hills rise steep and dark. Across the channel, Phillips Arm disappears into the mists. The feeling the place gives is brooding and mysterious. Two more boats in at last light; this is a favorite stopping spot. The boats, one of them a tug, belong to Americans, northbound like us, and the talk is of Alaska and the season ahead. Had a long look at that tug. Down in her basement was an old direct reversible Atlas engine—it must have stood 7 feet high, and just 120 hp. The skipper's worried, says he's got only 7 pounds of oil pressure. Think I'd be worried, too.

*APRIL 15*—Stumbled out at 4 a.m. A quick mug-up with last night's coffee, a word with Bruce, and down the channel in the black. Dawn, when it came, was delicate, pale yellow and red, with a few lacy clouds and the promise of another fine day. But it turned gray

*Sam resting with a paw draped protectively over his rock.*

after an hour, and the gusts reached us even in those narrow channels, rushing down from the hills and darkening the water. Greene Point Rapids, Wellbore Channel, Whirlpool Rapids. Missed the slack and tumbled through in the boiling ebb, the shores sweeping by close and broken water on all sides. Exhilarating for us, with a chart and with power to spare, but how about Captain Vancouver's crew, with oar and sail, exploring these passages in a longboat with no knowledge of the rips and overfalls in the passes?

Johnstone Strait at 11, with half a gale funneling through the hills of that windy canyon. The tide there runs like a river, and the shores are bold and steep, with snow close to the water even now. Fair tide today, so we stood in midchannel to take advantage of it. But we scooted for the beach at the turn, leaving the rip to a big tug and tow bucking south and taking solid water clear over the house. Johnstone Strait is a gloomy, inhospitable place, spring or fall, and I'm always glad to put it behind.

Bucked a fresh northerly all afternoon in Queen Charlotte Strait; later the wind switched around, just like that. With our decks running white, blinded by sun and spray, we beat our way up past that lonely stretch of shore, the *Kestrel* just a hundred yards away. Seas buried the bow twice, right back to the anchor winch, even though I had been chopping the throttle and letting the big ones hiss past. Once a green one stopped us dead, and I went below for a minute to

*Above—Bruce and Kathy's boat,* Kestrel, *north of Shoal Bay in Cordero Channel. Left—Susanna and Sam on the beach at Pruth Bay, Calvert Island.*

check the bilges. At 6 p.m. we reached God's Pocket, the tiny cove that's the jumping-off place for the most exposed stretch of water on the Inside Passage. The trees around the cove are bending in the gusts, the low clouds scudding by overhead, and spring seems far away. Just west of here are Nahwitti Bar and the stormy passage to the west coast of Vancouver Island—some of the wildest and least-frequented country on the whole coast.

Worked on deck tonight in a wool shirt and sweater, lashing everything down tight for the trip across the strait tomorrow. Forty long miles, and open to about everything. Bruce brought the *Kestrel* alongside. Susanna and I went aboard and we all went below to assess the damage of the afternoon. The vertical planking on their trunk cabin dries up over the winter, so the overhead is covered with ice-cream cups stuck in with fishhooks to catch the leaks, and the rug is wet to boot. But right now there's a fire in the stove and a jug on the table, so let it go. Bruce says he's all set for the morning, and I guess I am, too, but that's a nasty place out there even for a big boat, and I'll sure be glad to put it behind.

*APRIL 16*—Up at 3:30 a.m. to the sound of many engines starting. I looked out to a crowded float; five more boats came in during the night. A neighbor said the early report was fair, but poor for tomorrow. Off at 4, and already the running lights of the boats ahead were going out of sight in the big westerly swell. Glad for company today, and for just the palest trace of dawn before we got out into the strait. I don't care to face that stuff in the black. The day came unsettled with high clouds moving in from the southeast and a wind chop on top of the swell. Passed Pine Island and the keeper's house high above the white water and rocks. Made our turn for the long leg to Cape Calvert, ticking off the miles and watching the water uneasily for signs of what to expect.

The tug *Columbia* with a double tow passed us there, bound for the North Slope with a load of

about. The shores were broken and rocky, with the dark hills above and never a sign of man. Dropped the anchor with the *Kestrel* alongside in an empty cove, and went ashore to where a faint blaze on a tree marked an overgrown trail. Bruce, Kathy, Susanna, Sam and I made our way through thick dark woods, the trees like columns holding up the dark roof; even Sam was uneasy and stuck close to us. The forest thinned. From somewhere came the heavy beat of surf, and we stepped out onto a beach and stood there awed. Before us was a half-mile of white sand between high rocky bluffs. The heavy ground swell drove across the offshore reefs and boomed onto the sand. Almost a

*Above and right—The tug* Columbia *passes the northbound fish boats in Queen Charlotte Sound.*

housing for the pipeline crews. Day and night they pass now—tugs from all over the West Coast. Sleek new rigs and tired old wooden slabs with ancient engines, they slip by us.

It was an easy crossing on a raw day. Slipped inside again at Cape Calvert without a hitch and breathed a little easier, too, for I don't care much for big waters. Left the rest of the fleet at noon to slip off to the west, up a little-used side channel to a place I had heard

*Bunkhouses at the old Butedale cannery look out on Fraser Reach, Princess Royal Channel.*

dozen times I've been past this spot and never imagined that it existed. There were mink and deer tracks on the sand, and what I thought were wolf tracks, and all were fresh since the last tide. Sam ran and ran, barking in and out of the surf; we all walked with the sea in our ears and the sun and wind on our faces.

The night is chilly and starry—we're lying in this remote anchorage, perhaps 20 miles or more from the nearest human being. Susanna woke me in the night, when all was still, and said, "Listen." It was the far-away beat of the surf that she was listening to, beating on the shore across the island and carrying all the way to where we lay.

*APRIL 17*—Pulled the hook at 6 a.m., the day already warm and fair. Picked our way through the rock piles to Hakai Passage: tricky going with a heavy surge and white water clear across in places. I tried to get the *Kestrel* to lead, but no deal. Bruce says if I break my fiberglass boat, they'll give me another one free. Fitz Hugh Sound at 8, still and empty; Namu cannery, on the far shore, was the only spot of color on hundreds of miles of dark hills. Once there were dozens of these canneries in all the major bays and inlets, but now there are only ruins, and Namu is one of the few left on the coast.

Made our turn at Pointer Rocks, where the Inside Passage goes northwest through very narrow Lama Passage. Only a few miles to the east is the rock where Alexander Mackenzie painted his name in 1793, after coming through the mountain passes and down to the salt water from the east, the first white man to do so.

Passed one settlement all day, Bella Bella, at noon. The rest of the day was spent winding through still waters with the panorama of empty country opening up ahead and disappearing behind. The sky was an azure blue. We crossed Milbanke Sound and passed inside again to Tolmie Channel and Graham Reach, where deep canyons wind for miles into the interior between steep snow-covered hills with streams and waterfalls full of snow melt. Now and again side channels opened up, giving glimpses of narrow passages to remote inlets and bays that must see years pass without ever a visitor. Swanson Bay at 3. A single gaunt chimney was all that was left of the mill that supported small beach-logging operations along this section of coast 30 years ago.

Napped for a while. Ahead and behind us, the same strip of water between the high hills; never a boat, never a buoy, endlessly the same, hour after hour.

Tied to the old float at the Butedale cannery at 5, just as we do every year, and walked past row after row of empty buildings with snowdrifts on the north side, where the hills will cast shadows for another month. A town in itself, the cannery's empty now but for the caretaker and his wife, who sell a little fuel and groceries. The caretaker keeps the aging equipment greased and tends the great generators that are fed by the lake high above. I took the traditional Butedale shower, in the corner of a vast bunkhouse. There are rows and rows of empty rooms, each room with an iron bed, a chair and a bureau. On the walls were a few faded photographs of workers from seasons long past and forgotten. Clouds came over at dusk, heavy and ominous, and the night was very black.

*APRIL 18*—Sat up for a bit, early this morning, with a cup of coffee, listening to the roar of the falls, and watching the day start gray and rainy. If I get up early, I can sit and collect my thoughts before we take off. This place out in the middle of nowhere, so big and so empty, casts a spell over me each time I stop. The woods are slowly taking over now; I wonder how long it will be before we come around the point and the lights are out, the dam broken, the generator seized up, the place cold and deserted. Bruce started up with a puff of smoke, and the spell was broken.

McKay Reach, Wright Sound, Grenville Channel— what gloomy canyons they seemed today, with the clouds pressing down on the water, and the rain beating in heavy slants on our world of dark green hills and lead-gray water. The routine settled into ticking off the points as they appeared out of the mists ahead and disappeared behind. At 1 p.m. we swung into a little cove to watch a bear clamming on the beach. He would put a few clams into his big ham of a paw, smack it with the other, and lift the whole dripping mass up to his mouth.

Our moods pretty much follow the weather. Even our little marine radio was quiet as the dark shores marched past for hour after hour. We spent the

*Overleaf—Entrance to Grenville Channel, near Camp Point.*

afternoon with headwinds and foul tides, and dusk found us miles from the deserted cannery that was our destination. We gave it up and ducked into a little hole on the Lewis Island shore and dropped the hook quick. We settled into a tide-swept little gut, hard by the side of the mountain, just as the night closed in black and wet. Had dinner on the *Kestrel*—music on the radio, coffee on the boil; for a cozy spot with the dirty night outside, the *Kestrel* couldn't be beat. Bruce had the chart out, and we fingered it as the late Canadian weather came on: Wind from the southeast at 15 to 25 knots tomorrow, worse the next day. We have Dixon Entrance to cross tomorrow, another long open area, but we'd better go have a look.

*APRIL 19*—Up at 2 a.m. with a vicious squall blasting open the night and laying us over even in our sheltered spot. I fired up the radar and sat up to watch. Outside was nothing, not a light, not a star, only the pale picture on the scope showing we hadn't dragged. Went outside to let out more anchor line and double up the tie-up lines. The rain stung my cheeks and drove in under my oilskins. Off the stern the deck lights showed the trees looming close at hand. I went back to an uneasy sleep when the squall passed, the dog wedged between Susanna and me in the bunk.

But the dawn came clear with a high overblown sky, so off at 6 a.m. to Malacca Passage; then Chatham Sound opened up with the snowy wall of the mountains to the east. Ominous sky with a greenish cast to the southwest, but the noon weather was still clear, so we kept on, past Green Island and out into the open, with a heavy, long, westerly swell. The sun clouded over and the day grew very dark, but still there was no wind as we ticked off the miles. Cape Fox at 1 p.m. We breathed a little easier, with a harbor only two hours away.

When the wind came it was swift, darkening the water and blowing the tops off the swells, then driving long streaks of foam downwind. At first we thought it was just a squall, but then the announcer came over the radio with an emergency gale warning for the North Coast. We were at the last of the flood, and already the seas were getting steeper. With the turn, and the rush of the tide against the wind, those seas would just about stand on end. Bruce and I both jogged for a minute there, side by side, to put our trolling poles down for better stability; we secured our

hatches and got ready for what we knew must come. When the tide turned, we were a mile off a little rocky point, as lonely and forbidding a stretch of shore as there is on the coast. The tide race started on the beach and worked out; soon we were jogging dead slow in a jumble of white water.

So we came to Alaska, on a wild and lost afternoon, caught in a tide race off a nameless point, in failing light, far from any help. The heavy westerly swell, the dirty southwest chop, and the push of the tide on top, made it all I could do just to keep way on the boat, throttling over the big ones and then diving deep into the troughs. The seas came from all directions, and even at dead slow, waves slapped against the windows, sagging-in the thick glass. Twice a green one poured in over the stern, filling the trolling cockpit, and the boat wallowed deep in the water until it drained. *Kestrel* was just fifty yards away, and I could see half her keel as she came out of a big one. The shore wasn't far, and I looked long and hard at it. If the engine ever quit, we'd be broadside in a minute and probably swamp. If it came to that, I'd rather pour on the coal and put the bow in the trees than get off out here. Even a rocky beach is better to walk home on than this crooked piece of water. Bruce came on the radio: "I broke a spoke off my wheel on that last one"; I could feel the tension in his voice. I looked over and heard his engine change pitch as his bow lifted into another steep sea.

For three hours we jogged in that lonely spot, hardly making a yard. Twice it seemed to get worse, and for a time there was nothing I could do but try to avoid the worst of the waves and hope that everything held together and that no waves came through the windows. The light began to go from the sky with still no change; it was a desperate time. After dark, when a skipper has no way of seeing where the next wave is coming from, anything can happen. Bruce was off my quarter, and I looked over to him when I could, watching the seas bury his bow clear back to the cabin. We were climbing and diving into those smoky seas, and it felt pretty good to have a friend out there.

At very last light the push of the tide eased off, the seas seemed to lay down a bit, and we crept up the

*Heavy swell, tide and chop produce unpredictable big seas.*

*Ketchikan, and Snapper Carlson's packer* Westward.

beach again. We rounded the point in the black, with waves breaking heavily on reefs on both sides, and dropped the hook in the farthest corner of Foggy Bay, Alaska. Sam went up on the bow to sniff the air in that new spot as dogs do, and Susanna got up to have a quiet drink before we started clearing up the debris. The stove had blown out, so it was late before we ate and turned in. I think we went through a big one out there today. Susanna and the dog were in the bunk most of the time; I stuck my head down when I had a chance, and there were some pretty big eyes staring up at me. We had taken all the precautions we could, but sometimes you just get caught and there's not much you can do about it.

*APRIL 20*—Moved to the inner basin at first light, a snug spot where our rigging is almost in the trees, but with a view to the outside between the islands. The wind backed at 10 a.m. to a fresh southeast gale, and how it blew. For a few hours, outside was solid white, waves breaking 30 feet into the air over the reefs; I hoped no poor soul was caught out in that. Susanna's been finding things lost since our start; as for me, I settled in for fine lazy afternoon with a book. Sam was

restless to go ashore, but soon gave it up and slept against my leg.

*APRIL 21*—A rainy, gray day, but windless. Thirty-five easy miles to Ketchikan, where we tied up at 3 on this dreary cold afternoon, 10 days from Seattle, and snow on the hills close above. Dog disappeared before we even shut the engine down. Called and called, but no luck.

*APRIL 22*—Harbor day, with rain and the whine of the sawmill across the little basin. Bought supplies, did a few jobs of boat work, caught up on scuttlebutt. North tomorrow, for we don't fish this southern district much. Anyway, we're eager to get to our cabin near Point Baker. Still no sign of Sam, and a pretty quiet dinner we had, for that black dog is a big part of this family. Heard a bark just as we turned in, and opened the door to Sam, with scratches on his nose and one ear bitten clean through. Hope it was worth it, buddy.

*APRIL 23*—Left the mill and the town behind at 5 a.m., the morning clear and snappy—ice on Sam's water bowl—and the promise of the first fair day in a week. Overhead the big southbound jet passed, trailing smoke and noise, letting down for the airport miles to the south. Wondered what those people thought as they looked out of their little world to see two lone fishing boats headed out to the wide strait as the sun came over the mountain on another postcard morning.

Kashevarof Passage with rushing fair tide, Susanna and Sam asleep on the bunk as we swirled through with gulls and a porpoise for company. Sumner Strait, and the old stomping ground opened up to the west. A warm and fair spring afternoon. Parted with the *Kestrel* here. Laid alongside her for a long while, then she was gone, heading north to the cannery to drop off her nets and then out to the coast to troll.

These last miles were along familiar shores; the sky was red to the west, the hills black, the water still. Rounded the point at 11 in the cold and starry night, and picked up the old mooring off our own little cabin after more than five months away. In the stillness there was only the noise of the tide pouring around the point. The beacon on the rock flashed, but beyond that there was nothing.

*Across Sumner Strait, the Canadian Coast Mountains to the east.*

*"There are a few towns,
none . . . connected to
anyplace else or to
each other, except
by air or water."*

# Southeastern
# Alaska

A man in a boat could travel for weeks in Southeastern Alaska and never find a town. The coast, like the coast of British Columbia, is deeply indented with inlets winding back into a mountainous and forbidding interior. The several major islands, and countless smaller ones, form a maze of channels and passages between the ocean and the mainland. In the northern part of the region, glaciers lie at the head of most of the mainland inlets, and they discharge ice all year round.

A vast part of the area is thickly forested, without settlements or towns, little changed since the arrival of white men. Almost all the land is owned by the federal government in the form of national forests, and little is available to individuals.

There are a few towns, none large. Each has a few miles of roads, but except for Haines, in the far north, none of the towns are connected to anyplace else or to each other, except by air or water. The industries are

*Owl totem at Wrangell.*

fishing and logging, both fairly seasonal. Only the capital, Juneau, has a true year-round economy and escapes the seasonal nature of life in much of the area.

Tourism is now increasing, as many people travel through the area each summer on the ferries and cruise ships that pass almost daily during the season. But most of those people are going farther north or preparing to fly south. There is little to keep them in Southeastern Alaska, and I'm sure that after a while their trip just fades into a blur of islands, all the same, overcast days, and a few towns busy with canneries or sawmills.

Scattered in little coves and harbors far from the towns that are big enough to have fish plants, mills, schools and stores are a few roadless fishing communities that still enjoy a sleepy existence. One of these, Port Protection, was discovered and named by Captain George Vancouver when he anchored there on a gloomy and windy evening in September, 1793, after searching for a harbor in failing light and deteriorating weather. He named a nearby point and its small harbor Point Baker, after his lieutenant.

Today Port Protection and Point Baker each has a general store and homes built under the trees around the bay. All told, the population in the winter might reach 60 people. There are no roads, phones, electricity or other services. The nearest town is 50 miles away by plane or boat. Once a week a mail and freight boat makes a stop. The woods are thick, the homes are built hard by the water, and behind the homes the forest rises dark and tall. Even paths are few, and usually just used by dogs. For most people, skiffs are the way to get around.

Except for the storekeepers, all the men fish, mostly only for salmon. In the winter the men are all home; in the summer they're scattered from the bay out front to points up and down the whole coast. Their fleet ranges from skiffs to seiners, and when the fish run, it's a busy time.

*View from the cabin at Point Baker.*

*Scenes at Sumner Strait:
Right—In summer, the
floating cannery at
Port Protection.
Opposite—In winter, a
neighbor's float in the back
channel at Point Baker.*

Then the population of the area doubles and triples, as boats from the south and from other parts of Southeastern Alaska make one or the other of the two communities their base of operations for the season. Many boats have been going to Port Protection and Point Baker for years with their friends and sometimes with their families; when the fishing period is over, or in the evenings, the floats of the communities are almost towns in themselves.

But it's a rush too, for the peak of the run may pass the area in just a few days, and when the fish run, it seems as if there aren't enough hours in the day for everything that you have to do.

But then comes the fall; the outside boats leave, and the local boats straggle back home. By the first of November, everyone is pretty much tied up for the winter. In the winter the days are short, with a pale sun low in the sky. Many homes are hidden by the trees and don't feel the sun from November until March. It isn't a bitter winter—rarely will the bay freeze—but the weather is poor and the snow lies on the ground for weeks at a time. In the summer, as many as two and three planes a day land in the bay, bringing people and supplies, but in the winter it's dead quiet, and weeks will pass with only the mail boat to break the monotony.

Despite the short days and gloomy weather, many of the local residents prefer the winter. Salmon season's a rush and the winter can be a welcome change, with time to work on a cabin or boat, visit neighbors or just sit inside when the weather's poor. It's not a fast life, but there's enough to do just to keep going. Many of the residents have spent time in the larger towns and wouldn't think of ever going back.

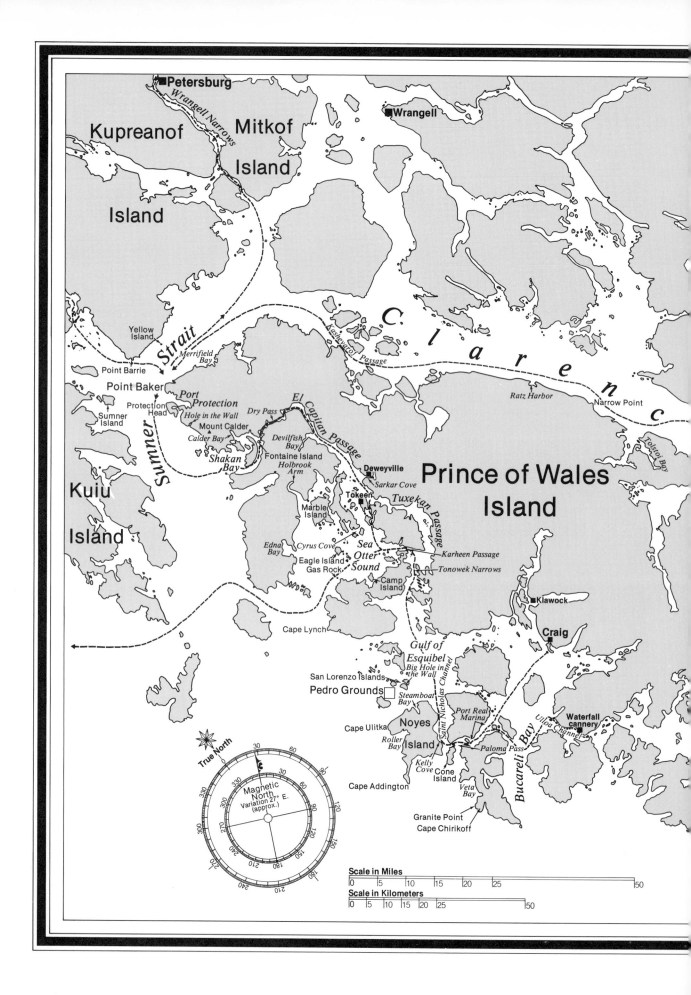

■Petersburg

Kupreanof

*Wrangell Narrows*

■Wrangell

Mitkof

Island

Island

*C l a r e n c*

Yellow
Island

*Strait*

*Merrifield
Bay*

*Kashevarof Passage*

Point Barrie

Point Baker

*Ratz Harbor*

Narrow Point

*Port
Protection*

Protection
Head

*Hole in the Wall*

*Dry Pass*

*El Capitan Passage*

Sumner
Island

Mount Calder

*Sumner*

*Calder Bay*

*Devilfish
Bay*

*Tolstoi Bay*

*Shakan
Bay*

Fontaine Island

■Deweyville

*Holbrook
Arm*

*Sarkar Cove*

## Prince of Wales
## Island

Kuiu

Tokeen

*Tuxekan Passage*

Island

Marble
Island

*Edna
Bay*

Cyrus Cove

*Sea
Otter
Sound*

*Karheen Passage*

Eagle Island
Gas Rock

*Tonowek Narrows*

Camp
Island

■Klawock

Cape Lynch

*Gulf of
Esquibel*

■Craig

*Big Hole in
the Wall*

San Lorenzo Islands

Pedro Grounds

*Steamboat
Bay*

*Saint Nicholas Channel*

*Port Real
Marina*

*Ulloa Channel*

**Waterfall
cannery**

Cape Ulitka

Noyes

*Roller
Bay*

Island

*Paloma Pass*

*Bucareli Bay*

True North

*Kelly
Cove* Cone
Island

Cape Addington

*Veta
Bay*

Magnetic
North
Variation 27° E.
(approx.)

Granite Point
Cape Chirikoff

Scale in Miles

| 0 | 5 | 10 | 15 | 20 | 25 | | 50 |

Scale in Kilometers

| 0 | 5 | 10 | 15 | 20 | 25 | 50 |

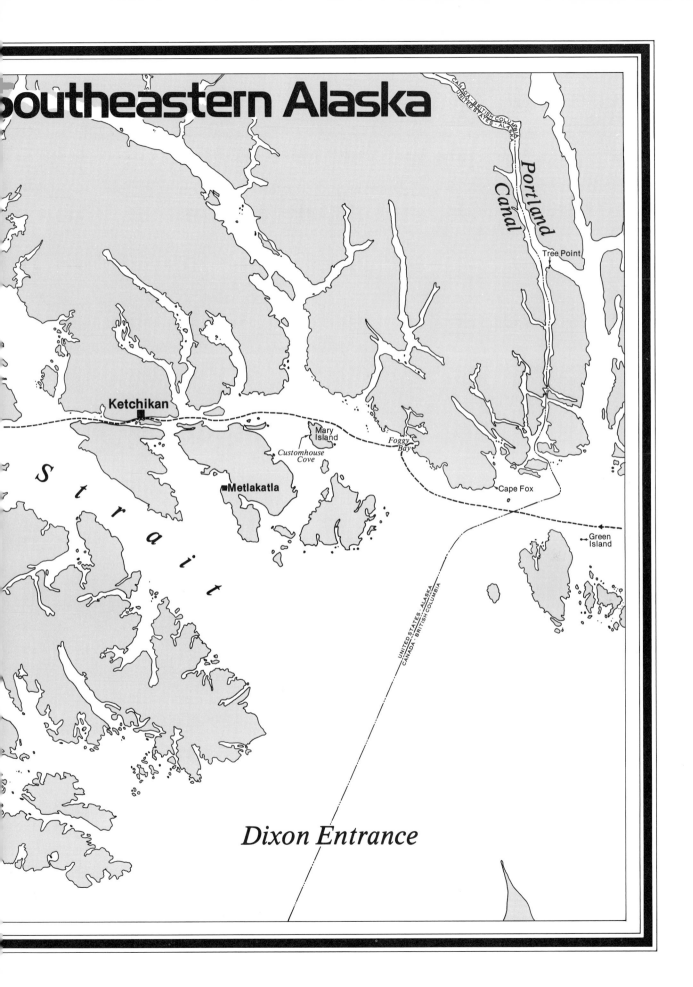

# Southeastern Alaska

Portland Canal

Tree Point

Ketchikan

Mary
Island

*Customhouse
Cove*

Foggy
Bay

■Metlakatla

Cape Fox

←Green
Island

CANADA - BRITISH COLUMBIA
UNITED STATES - ALASKA

*Strait*

UNITED STATES, ALASKA
CANADA - BRITISH COLUMBIA

*Dixon Entrance*

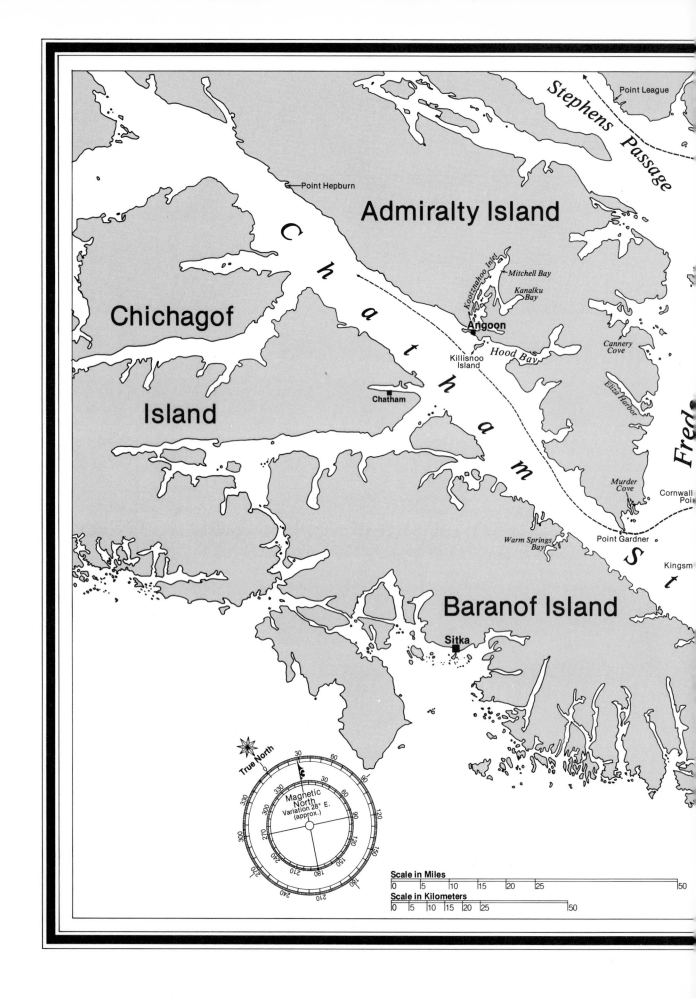

Stephens Passage

Point League

Point Hepburn

**Admiralty Island**

Kootznahoo Inlet

Mitchell Bay

Kanalku Bay

**Angoon**

Killisnoo Island

Hood Bay

Cannery Cove

**Chichagof**

Eliza Harbor

**Island**

Chatham

Fred.

Murder Cove

Cornwall Poi

Point Gardner

Warm Springs Bay

Kingsm

St

**Baranof Island**

**Sitka**

True North

30

60

90

30

60

Magnetic North

Variation 28° E. (approx.)

330

300

270

240

210

180

150

120

90

Scale in Miles

0    5    10    15    20    25                    50

Scale in Kilometers

0    5   10  15  20  25                50

# Southeastern Alaska

CANADA · BRITISH COLUMBIA
UNITED STATES · ALASKA

Stikine River

Petersburg

Wrangell

Kupreanof

Mitkof

Wrangell Narrows

Island

Island

Keku

Strait

The Summit

Rocky Pass

Devils Elbow

lets

rity

hington

Kuiu

Yellow
Island

Merrifield
Bay

Clarence Strait

Kashevarof Passage

Summer

Point Barrie

Point Baker

Strait

Port
Protection

Dry Pass

El
Capitan
Passage

Prince of Wales

Bay of
Pillars

Protection
Head

Hole in the Wall

Sumner
Island

Mount Calder

Devilfish
Bay

Island

Tebenkof Bay

Calder Bay

Shakan
Bay

Fontaine Island
Holbrook
Arm

Deweyville

Sarkar Cove

Tuxekan Passage

Island

Port
Malmesbury

Marble
Island

Tokeen

i

t

ort Walter

Port Walter

Edna
Bay

Cyrus Cove

Eagle Island
Gas Rock

Sea
Otter
Sound

Karheen Passage

Tonowek Narrows

Camp
Island

Port Conclusion

Port Alexander

Cape Lynch

Gulf of
Esquibel

Big Hole
in the Wall

Cape Ommaney

San Lorenzo Islands

Pedro Grounds

44

*Susanna in her garden. The net may, or may not, keep the dog and the wildlife out.*

APRIL 24—A peach of a day, fresh and clear with light northerlies. Sam couldn't wait this morning, so over the side he went, to swim ashore to the point and run through the woods barking until we went ashore, too. We have the whole cove to ourselves. A half-mile away, through a narrow tidal channel, is the settlement of Point Baker, but here there's hardly a sign of man. The cabin we built a few summers ago is high on a point with water on three sides, but hidden from view by the trees. It's just 12 by 16 feet with a sleeping loft above and storage out back, but that's fine for the three of us. I pulled the boards off the windows and laid a fire as Susanna took off through the woods to check on her garden spot. Flying squirrels had found their way into the cabin and chewed on the food, but otherwise all was tight and dry. Spent day in unloading the boat and just settling in. I sat for a bit in the evening with the sun dying brilliant in the hills across the strait, our boat riding easily in the cove below, Sam asleep at my feet, and Susanna beside me. If there are any fish at all, we'll just stay here and fish.

APRIL 25—Fair again. Spent the day in the garden with Susanna and Sam. Spring's late up here, but already the days are getting longer and the first buds are appearing, so we planted anyway and strung up some old net to keep the dog out. If we have to go out to the coast, it'll be too late to plant by the time we get back.

A neighbor came by tonight, said not to waste our time fishing here; most of the local boats left last week for the coast. Guess we'll have to go, too, but don't much want to. Built this place so we could get off the boat once in a while, and every year since, we've had to travel farther and farther to fish.

APRIL 26—Wind southeast at 25 knots with spits of rain. Spent the day in boat work: changed the oil and went over the trolling gear, made all ready to travel in the morning. Back to the cabin, where I found Susanna had baked rolls, bread and even two pies for the trip, but I don't think she wants to go any more than I do. Sam fell down from the loft last night. He climbs up the ladder to sleep with us, then gets hot in the night and circles around for a cooler spot. Heard a great crash and looked around with the flashlight. Finally spotted him asleep under the table downstairs; he probably fell seven feet and never even woke.

*Troller, cabin and float at Point Baker.*

*APRIL 27*—Weather continued southeasterly with squalls. Gathered up wife and dog and tried for early start, but ran into a friend while taking water at Port Protection, and spent the morning drinking coffee and catching up on the winter's gossip. Missed the tide and spent a dirty afternoon bucking down Sumner Strait toward the ocean. The outside route looked gray and mean, so we headed up Shakan Bay and back into the trees and rockpiles of El Capitan Passage. It's a longer and more tedious route, but it puts off that outside business for a while longer. This time of year it's still almost winter out in the Gulf of Alaska, and I take the sheltered routes every chance I get. Ran through Dry Pass at the very last light, through channels with no room even for two boats to pass, the trees almost meeting above. At 11 p.m. slipped past the rock at the entrance to Devilfish Bay and dropped the hook in five fathoms, soft bottom, the night black and rainy. Shut down to the roar of falls close at hand; above us,

pale snow on the steep hills. Untraveled country out here; passed only one boat all day.

*APRIL 28*—The day came overcast and found us swinging at anchor in a little landlocked bay with trees rising steeply to snow on all sides. In the shallows ducks traded back and forth, and off our stern three seals played, diving and swimming under the boat. Soon we'll be in the outer islands, with the ocean close around the point, and such peace will be rare, so we savored it this morning, dawdling over breakfast, Susanna looking through the glass at the ducks with her bird book out. But miles to go, so we went down the long channel at 10, while watching the depth recorder for signs of fish as we passed into island-choked Sea Otter Sound. The country grew ever wilder and more remote, with steep-sided islands and lonely inlets, and only an occasional logged-over slope to show the hand of man. Ran all the old spots with the sounding machine, but not a sign of fish or feed, either. Anything at all and I would have stayed, but it just looked dead. Saw one boat and he was running, too, so I guess it's the big waters for sure; can't put it off any longer.

Passed the ruins of a nameless settlement in narrow Tuxekan Passage, already the wind darkening the water to the southwest. The Gulf of Esquibel opened up ahead for 20 miles of rolling in the trough, with bleak dark islands, gnarled trees and forbidding rocky shores on all sides. The closer we get to the coast, the wilder the country gets. A rain squall passed over us, blotting out the shores and causing squealing on the radio. Finally, at 5 p.m., the fjordlike opening of Steamboat Bay on Noyes Island appeared out of the mists, and we passed inside to tie up, the fourth boat out at the float of a disused cannery. Vicious williwaws buffeted us from the mountains above; a neighbor says no one has left the bay in three days.

Baited up and laid out all my gear for the morning anyway, while Susanna and Sam headed off for the woods. She beat me four hands straight at cards after dinner. As I write this, in the stillness between the gusts I think I can hear the boom of the surf on the outer coast of the island three miles away.

*APRIL 29*—Day came at 3 a.m., with violent squalls heeling us over at the float. Pulled on oilskins and ran a long bow line out to the float, then back in that sack for some good winks with the eerie whine of the wind loud around us. I'd rather be in here than swinging on the hook somewhere and worrying about dragging. Slept late, and no one even started an engine; you can lie in your bunk right here and listen to the weather report howling in the rigging. Fed the minks last night, I guess—not a bait left on the hooks this morning. You're falling down on the job, Sam.

Out in the wind and rain for a long walk with Susanna and Sam. Sam's good company these long harbor days, and he gets us out of the boat. Lying in the bunk rainy day after rainy day, with the wind shaking the rigging, can get to you after a while. Walked past rows of bunkhouses and outbuildings,

*Lying in and waiting for a break in the weather; Steamboat Bay, Noyes Island.*

48

their paint peeling and windows broken. Huge scows for fish trapping, with steam winches to haul the heavy trap anchors, were lying on the beach, rotting and broken. Around the curve of the beach we walked, until the cannery seemed tiny. Later, dinner and more cards, with wet clothes drying over the stove. That cabin back at Baker looks pretty good tonight.

*APRIL 30*—Gray and cold, with engines starting up on all sides at 5 a.m. I put on two wool shirts and followed the *Anatevka*, skippered by Doug, a fishing friend of seasons past. Headed out to Cape Ulitka, where the ocean pours around the point and meets the outgoing tide. That's a dirty spot; we almost turned around right there. The swells on the outside were 15

*Left—The cannery at Steamboat Bay.*

feet high, mean and ominous in the early morning light with a wind chop on top. Ran out to the 30-fathom edge and went back into the trolling cockpit to start dropping my gear. A mile and a half away loomed the dark wall of Noyes Island, clouds low on the hills and the shores lost in the spume and mists of the heavy breaking seas. They call this the Gulf of Alaska, but it's the North Pacific Ocean, and the shores show the effects of the winter gales: bare rock extends up over a hundred feet from the water without a trace of vegetation, and above that what trees there are grow twisted and stunted from the constant gales. It's a wild spot to fish, but when the fish come by, a fellow can do well even in a small boat—if the weather will let him fish. They're traveling fish, running south along the Pacific Rim and headed for all the major rivers up and down the West Coast; I doubt that one

in five is headed for an Alaskan stream. We snag a few on the way by. My boat is well built and I trust it, but those are big waters out there, and I fish with an eye over my shoulder, for the weather can blow up in a hurry, and then getting through the rip at the cape can be a desperate struggle.

First day of the season, and the kid is all thumbs. Always there are the little bugs to work out after the winter and working on gear, but today I lost a whole line—lead, flashers and all, cut off by the stabilizer dragging back too far. Susanna and Sam pretty much stayed in the bunk, so I was back in the cockpit trying to hang on, fish, and steer the drag by memory.

I was disgusted by 3 p.m., and the weather was doing things to boot, with spitting rain and a freshening breeze to the southwest. Looked around to see the *Anatevka* close, but no other boats in sight. We crossed tacks just then and I waved that I was picking up; Doug was picking up, too. By then the tide had carried us miles to the south and we figured we might as well just go around the island.

If you only see Cape Addington once, it should be on such a day. The wind was coming on hard and running against the tide, and the cape, a long rocky arm, was almost lost in the mists as the heavy seas beat against its rocky sides. That's an evil place, and we gave it plenty of room in order to stay clear of the tide race, but even so, it was a wild hour before we were around. Once, on top of a big one, I took a long look around. To the south the water was white and the sky dark. A

Anatevka *trolls through the rips at Cape Addington.*

*Trying to put the gaff into a fish.*

quarter-mile away, Doug labored heavily through the rip, but beyond that we were alone. There was nothing but the sky and the troubled sea. We rounded the cape, finally, and after that the rest seemed easy—an hour's run along a bold shore, then in through the steep swells at Cone Island to tie to the little float at Kelly Cove on Noyes Island. The crew rowed ashore, and I poured out a stiff drink and started in on the mess on the back deck. Seven fish for the day—just about covered our losses.

*MAY 1*—Up at 5 a.m. to a pale clear morning, not a tree moving around the cove. Under way at once, but I didn't trust the weather much. Heavy swells at the entrance; idled through with the tide pushing us on. Fished Veta Bay and Granite Point and had a dozen by noon, when it blew up southeast and we had to give up. Had a good long walk on beach this windy afternoon, though. Susanna says she's going to Hawaii this time next year. Don't blame her much. Haven't met too many people that really like Noyes Island in the spring. Don't like it myself, but it pays the bills.

*MAY 2*—Actually fished a full day—Cape Chirikoff for 18, and most were large kings at $1.20 a pound! Gear, finally, more or less fishing well.

Weather continued southeast. Lousy but fishable. It's exciting to get into a bite of kings—king salmon—and maybe get two or three on a line. Had a couple like that today, but mostly it's just standing back there in the cockpit, wedged in, watching your depth on the meter and watching the lines for a hit. Lost what looked like a 40-pounder today—dammit—had him right up to the boat, ready to lay that gaff alongside his head, and he just took off and broke the leader. Pulled up three useless hooks today—solid duranickel and bent back straight like coat-hanger wire. Should have bought stainless steel. Sometimes I think that if you land 50 percent of the fish you hook, you're doing pretty good. But God, it hurts to lose those big babies—that one today was a $50 bill.

*MAY 3*—Couldn't believe it—bright sunshine and no wind until the afternoon westerly; can this be Noyes Island? Fifteen fish, too, and Susanna even went back there with me for a couple of hours and landed five—how do you like that? Then, on the way in, found a big glass float, covered with weeds and kelp, but unbroken. Wrote in the log book in capital letters: BIG DAY. Topped it all off with a hot shower at the buying station in the cove.

*Kelly Cove, a 500-foot-wide safe harbor on the southeast corner of Noyes Island only a few miles from Cape Addington. Seen here with fish boats around it is the cove's big fish-buying scow, a double-decker with a store and showers above and floats on both sides.*

*"Trolling, especially for king salmon,*
*is close to being an art. . . .*
*The color of the sky, the time of day,*
*the flow of the tide:*
*all these are important. . . ."*

# Trolling for Salmon

Trolling for salmon is a hook-and-line fishery—very different from net fishing. A boat will fish four stainless-steel lines that are reeled in and out on individual gurdies (winches) powered by the boat's engine. At the bottom of each line is a weight heavy enough to make the line lead straight down as it runs through the water. The actual fishing gear, the leaders and lures (or baits), is attached to the line at intervals and trails in the water. This way, a boat with four lines may be fishing up to 30 or 40 individual pieces of gear.

The tall trolling poles (carried upright on each side of the boat and lowered before fishing) provide means of deploying four lines: The two lines streamed from the tips of the poles (one from each) provide maximum spread of the gear. Lines from the midpoints of the poles trail astern to floats perhaps 30 yards behind the boat, and from the floats straight down into the water. Bells on the poles signal when a fish is on a

54

*Above—Kestrel, trolling
poles stowed, in
El Capitan Passage.
Right—Trolling gear; the
lures are coho spoons.*

line, although a large fish will often shake the rigging so violently that a bell isn't necessary. To haul in a fish, the fisherman engages the appropriate gurdy and reels in the line. As the gear comes up to the boat, each leader is unsnapped and coiled on the stern until the one with the fish comes up. The trick then is to pull in the leader (6 to 18 feet long) by hand until the fish is close enough to gaff aboard. But pulling a fighting 50-pound salmon close enough to gaff is easier said than done, and many fish are lost in the process.

Trolling, especially for king salmon, is close to being an art. Kings bite on different gear in different areas at different times. The color of the sky, the time of day, the flow of the tide: all these are important to the troller, and a man might spend a whole season just getting to know the ins and outs of one bay. King salmon trolling means knowing the bottom and being able to keep the gear close to it, but without snagging and ripping the gear. Kings feed mostly on the bottom. They often prefer rockpiles that only the cleverest trollers can fish.

*Skiff trollers in Sumner Strait.*

Fishing for coho, or silver, salmon is easier—the fish swim off the bottom and are less choosy about what they will bite on. But they are smaller and worth less: 50 kings would be a super day, but 50 silvers wouldn't be that much. Generally, trolling for king salmon takes place in the spring and early summer. Then it tapers off about the time that the heavy runs of silvers start, and fishing for silvers continues into the fall.

Few areas are closed to trolling; essentially the whole coast and most of the inside waters are open from spring until fall, seven days a week.

The trolling fleet is mostly old, traditional boats, built more for their sea-keeping abilities than for comfort. Long, deep and narrow, they will often be able to run in the trough of a sea that a modern, wide, hard-chined boat will have to quarter away from.

Some trollers fish in only one area, just putting in their time, good days and bad, until at the end they have a season. Others fill their holds with ice and travel. They might make their money in one bay or they might make it in a dozen. It is possible to put in a trolling season and for the most part fish by yourself, with few boats around.

Perhaps more than any other salmon fishery, trolling requires great patience. Day after day, a man will plug away, only making wages. Then he'll pull in fish two and three at a time, and maybe make a good part of his season in just a few days.

*M*AY 4—Southeast wind, naturally. Got an early start and stayed until 11 a.m. for 10 fish, but the wind was too much and we headed in; the big boats weren't far behind. Very big swell at Cone Island; running against the tide, we couldn't keep up with the big seas even at 2,200 rpm. Overtaken by a very steep cresting swell, but it passed underneath and didn't break. Don't think I want that for a steady diet.

The fleet's all in here at Kelly Cove; much socializing in the afternoon. It's just a tiny watchpocket of a cove, but there's a big buying scow anchored here, a double-decker with store and showers above and floats on both sides. Many of the boats put in their whole season right here, so it's a pretty congenial spot. Doug came over for dinner tonight. He's fishing alone, and I don't know how he does it. I have a hard enough time, and there are three of us.

*MAY 5*—Up at 5. Sniffed the wind, fired up, and went out through the swells at the entrance. So it goes. The day was gray with spits of rain, and the weather report poor, so we stayed at Granite. Had only 14, but a

couple of them must run to 30 pounds or better, so that's a good day. There must be 50 boats out there some days, but, with the big swells and all, they just get lost; when I look around on top of a sea, I'm lucky if I can see half a dozen.

*MAY 6*—Queer business this morning—we were backing away from the float with Doug behind when he went dead in the water. He called and said that something had killed his engine and that he was taking water in the stern. He got going O.K., but was still leaking and had a thump coming from around the shaft, so he unloaded at the scow and took off for town. Could have been us just as easily. The weather was still poor; fished Granite again for 10, but sure missed something big—it shook the rigging and dragged the line way back, but when I hauled in, the fish had taken the leader with him—120-pound test.

*MAY 7*—Clearing with a southwest wind at 20 knots, the last day for this trip. Tried trolling in the deeps today off Chirikoff, having seen some pretty good boats out there these last days, but dragged down to 70 fathoms all morning for just a handful—all big, though. I hauled all the lines back at 3. Susanna handed me a coffee with a shot in it, and we headed for

town with our first fair trip. We could have sold our fish at the scow every day instead of icing them all in the hold. But if we take them to town, we get a few more cents a pound, and the trip gives us a break from the routine and a chance to buy new gear if we need it. The sun even came out as I cleaned up the deck, and I actually lazed around for a few mintues with a beer and watched the town of Craig appear out of the dark hills in the distance.

Sold the fish, iced and fueled the boat, and had all the work done by 6, so we headed uptown for dinner in this sleepy Indian town. Sat in the restaurant looking out at the water and the evening sky. Spring fishing on the coast is a difficult and worrisome business, and sometimes it's hard to make a living and pay attention to your family too. So now and then Susanna and I just stop for a day and get completely away from fishing, even if it means losing a dollar or two.

*MAY 8*—Cool gray morning. Slept in—first time in a week. Then across Bucareli Bay to where Ulloa Channel opens in the hills and the Waterfall cannery sprawls along the steep shore. We had heard about the deserted cannery, but still were unprepared for the sight, and just looked at it in awe. It was almost a city,

*Top left—Trolling poles down and working, Lazy J bucks into a sea off Granite Point. Left—Waterfall cannery. Overleaf—Piles from a one-time dock at the old cannery stand in Ulloa Channel.*

*Company housing at Waterfall cannery. The cannery was once the largest in Alaska.*

stretching for a mile along the channel. Row after row of white buildings and docks and huge warehouses, eerie, the paint peeling, the windows all boarded up. It used to be the biggest cannery in Alaska, they say, but it was disused, vacant and for sale when we were there. It has since been made into a sport fishing resort. These days, fast, refrigerated fish-packers take the salmon to town, where the labor is cheap. But that's only a part of the reason there are so many deserted canneries. The other part is that the big runs of fish that the canneries were built for are gone now, fished out, perhaps forever.

The day was chill, the sky gray; we spent the afternoon walking deserted streets between empty and silent buildings. Some of the buildings are three stories high and hundreds of feet long. Inside, the machinery was all gone, acres of floor left between the rows of windows. In the slough behind were trap scows, pile-drivers and huge barges, all drawn up on the banks and rotting, with trees growing up through the hulls. With the breeze cold off the water, and the dark trees and hills silent behind, it was depressing, and we were glad to get back to our cozy boat, and leave that gloomy place behind.

Ran north, then west, through Port Real Marina, with squalls driving down from the hills and a gale forecast for the night. Dusk came quickly and the night was inky black, with not a light anywhere. Tried anchoring in a bight on the south shore, but we were blown out by squalls twice, so we went through Paloma Pass, with the tide running, hardly a hundred feet

between the reefs—and not a light or marker anywhere. Some boats won't go through Paloma Pass even in daylight. Spotlighted our way through and anchored right in the trees on the other side, in seven fathoms, mud bottom. As I write this, Susanna has dinner on the stove, Sam is asleep against my leg, and outside the wind rushes in the tops of the trees, but little reaches in here.

*MAY 9*—At anchor, Paloma Pass. Today the seas drove white around the point and beat on the rocks off our stern, but in the lee we were secure, our world bounded by the dark shores and the racing clouds above. Neither on this island nor most of the islands around is there any sign of man. Saw one boat yesterday, none today; tried to get music on the radio but there was only the hiss of static. So our isolation is pretty complete. But it's cozy enough in here; sat up with a fresh pot of coffee while I tied up a few leaders. Susanna read in the bunk with Sam snoring at her side, and after a bit I took a nap myself. Up at 3 to a new howl in the wind as the gale backed farther around and blew even stronger than before. Outside the seas were wild, and the trees bent with the force of the wind. Pork roast for dinner tonight, with spuds, onions and applesauce, and fresh pie for dessert—all

out of a tiny oil stove. Just like home here, even though the night is falling on a wild and empty coast outside.

*MAY 10*—The day came with low clouds and spits of rain, but the wind was just 15 knots southeast in the anchorage, so we went off to have a look at the passage out to the ocean. The water was breaking all the way across, so I gave it up and headed up Saint Nicholas Channel. The coves and bights on both sides were full of boats. At 3 we reached Steamboat Bay, where Sam was off for the woods before we even got tied up. Had a fine long shower in the cannery this afternoon. There was lots of hot water, and I found an old piece of soap with plenty of miles left on it. The lights didn't work, so I sang in the dark. A mail plane came in on its third try, doing some fancy steps on the water between the gusts. If it were me, I believe I'd tie that plane up and wait for better days. The pilot took off soon after he came, and was immediately lost to sight in the gloom and low scudding clouds. We baited up and went to bed early.

*MAY 11*—Up in the pale predawn at 4:30 to hear the early lies from the weather service and see lights winking on in the boats up and down the float. The procession started and we were right behind. Cape Ulitka at 6, the swells as big as ever, but the day was fair with light airs from the southwest, the clouds having been blown away in the night. Fished those watery hills—one moment we'd be down in the trough looking up at their green sides, then we'd pop out on top with breakers to the east and nothing to the west between us and Japan. But the watery hills were only swells, and what a day we had out there—bright sun, and an entire hour with a fish caught every few minutes, Sam on the hatch barking as they came over the side. Thirty for the day and most of them large—our best day yet. A week like this and I might be able to write a check again.

Talked to Bob tonight on the big radio. Heard him for the past two nights, calling us from down in Canada somewhere, but he was too far away and we couldn't get back to him. Tonight he just boomed in, and we talked for a few minutes until he faded out and I lost him again. Here we are in the space age and sometimes that radio still amazes me. He's anchored in some lonely hole in Canada, a couple hundred miles away, and for a bit it was just as if he was next door.

*MAY 12*—Windy with rain from the southeast. Tried the Pedro Grounds, an inside drag, for a change. It's a rocky bottom, and you have to watch your depth meter pretty closely the first couple of passes, but still,

you can catch a fish there when it's too tough on the outside. Susanna is pretty good at running a drag like that, so I turned the whole business over to her after lunch and took a nap. When I woke she had half a dozen good-sized fish.

Laid at Big Hole in the Wall, a tiny cove on the easterly island of the San Lorenzo Islands with another buying scow and float. On the beach a row of rotting shacks attested to better days. Counted 30 of us in here after dark tonight. We weren't the smallest boat, but there sure weren't many smaller.

*MAY 13*—Outside again for 26 off Roller Bay. That's almost too far out to fish by landmarks on shore, but—if I can make out where two hills line up

*Trolling on the outside of Noyes Island: Above*—White Cloud, *four lines working, at Cape Ulitka. Bottom right*—Gurdies and lines off *the stern quarter of the* Doreen *at Roller Bay.*

on shore, and if I graze the bottom with my leads —there's a spot where I'll almost always come up with one or two fish. That's something, to put your hand on the line where it goes over the side, feel it as it grazes the bottom with a soft twanging, run the range, and then, if you're lucky, wham! a strike. Weather southeast on and off, mostly on. Felt pretty good until I learned the top boat at the float had 53.

*MAY 14*—Blown in at 2 p.m. with just 10 for the day. Started catching a bunch of small fish, too small

to keep, and moved to deeper water. Hate to shake those little guys off the hook, because you know most of them will die. Think we'll keep running into Steamboat Bay (Noyes Island) in the evenings. Big Hole in the Wall's really closer, but there's no place to go ashore and walk, and some days I don't know who needs a walk more: us or Sam. Asleep tonight with the usual whine of wind in the rigging.

*MAY 15*—Went out into the southeast swell and chop this morning; pretty nasty stuff, but went out anyway. Shouldn't have done it, either—plugged up a fuel filter out there, and just about rolled the rails under before I got the filter changed and things going again. I did some pretty fast wrench work. Lines tangled badly, but we didn't lose anything. Twenty-four fish for the day, and we earned them.

*MAY 16*—Same spot, same weather, more fish—34—our best day yet. But that weather out there is beginning to get to me. Why can't we have good fishing with good weather, for a change? Some days something rises within me out there, jogging through that rip at Cape Addington. I push it down each time, but it's getting stronger. Sold fish at Steamboat tonight; can't afford to take a trip to town and maybe miss a day with the fishing like this. Very strong wind again after dark; doubled up the lines before turning in.

*MAY 17*—We were blown in today; the weather was too tough even for fishing the inside. No one left the float. Good thing, too—if someone goes out to have a look, everybody else gets twitchy, even if it's blowing a hurricane. Talked to old Henry today, in the cozy little cabin of his converted 26-foot whaleboat. He's rigged for sail and he has a little one-cylinder diesel for power. Last year he took off from here in August, right out into the ocean, and he sailed day and night seven days until he got down to Neah Bay, in Washington. He said this spring the weather's too much for him.

*MAY 18*—Nasty day, with half a gale from the southeast, but everyone else went out and I sure wasn't going to stay behind. Seventeen fish for the day, but I think we must have lost twice that number, jerked right off the hook by the violent motion of the boat in the sea. Saw a 300-foot Japanese stern trawler out there, just cruising north along the 12-mile limit (it's 200 now). Awesome sight, that—there we were just

rolling our guts out, and that guy was running at about 20 knots, rock steady. Signs of mutiny in the crew today, and I don't blame them much. In the worst of today's weather I just about had to crawl up to the pilothouse. The *Kestrel* called—they're running with Bob, and headed this way, but the weather blew up and turned them around at Cape Lynch. Said they'd try Sea Otter Sound for a day or so. Good; maybe they'll find something there and we can give up this outside business for a while.

*MAY 19*—Winds southeast at 25 knots, but fishable. Spent a dirty morning out there, with a big wind chop on top of the swell. Think Susanna started to say something when we made the turn toward Cape Ulitka, but thought better of it. Just a handful of fish this morning, and weather deteriorating all the time, so I fished with an eye over my shoulder. We were quartering away from the waves, but I still had to turn and put my stern to the big ones, and even so, I half filled my cockpit once. Bob called again at noon; they were inside Sea Otter Sound and getting a few. Inside waters sounded pretty good to me, and I didn't need much coaxing to pick up.

Took a long look around as I pulled in the gear. Inshore there was another boat, fishing the very edge of the breakers and almost lost in the swells; beyond that we were pretty much alone. The wind was coming on hard and driving long streaks of foam down the face of the seas. We had a dirty time again at the cape, but finally passed inside, and I gave the wheel to Susanna and went out to clean the deck with a thoughtful eye on the broken water behind.

I didn't think we'd be that way again for a while, not if there were any fish at all on the inside. We had put in our time on the coast, and did well; caught our share, and the boat never let me down. But it's a worry out there for a little boat, fishing alone, or even with a partner. Those are big waters and it's a worry, make no mistake about it. So we were pretty glad to leave the windy coast this afternoon and try our luck on the inside for a while.

Crossed the Gulf of Esquibel under lowering skies and a rising southeast wind. Tonowek Narrows and Karheen Passage. Then the night settled in, inky black; not a light; not a star. At 10 p.m. we laid alongside the *Cape Hason* at the log boom on Camp

Island in Sea Otter Sound. The *Kestrel* was there, too, and we all crowded into Bob's cabin aboard the *Cape Hason*. A winter had passed since the last time we were all together. Then, we were holed up in Canada somewhere with fall gales marching by on the outside. Now we were here, with all of Sea Otter Sound to ourselves and the whole season ahead of us. Fish, watch out! Bob's wife put their two little girls to bed, and we talked until deep in the night.

*MAY 20*—Up at 5 to run a mile through glassy waters and dump the gear in just as the pale day broke in the east. Smooth bottom, an easy breeze, and sun; even had one hit as I was eating breakfast. This sure makes the coast look sick.

Sea Otter Sound is roughly 10 miles by 12, almost landlocked, and sheltered from most weather. With many islands and little holes to fish, intricate shores,

*There is a great difference between fishing the outside, such as out off Noyes Island, and fishing protected waters in among the southern district's numerous small islands: Left—The rockpiles off Noyes Island. Below—The back channels and passages of Sea Otter Sound, flat and placid.*

and a few boats around, it's the kind of fishing we dream about.

The day was fine, and the fishing was good. I fished only that one drag, back and forth, sharing it with Bob and Bruce, holding up our fish as we crossed tacks. Almost gave it up at 7 in the evening, but had a little clatter and kept going until dusk. The stars were coming out by the time we pulled our gear and headed in. Seventeen for the day, too. How good it felt to catch a few fish and not beat ourselves to death doing it.

*MAY 21*—The day came overcast, but still there was a warmth in that thin sun that we didn't get out there in the ocean. Harley and Chuck, friends from past fishing seasons, showed up, but aside from them we had the whole sound pretty much to ourselves. Eleven fish today, and it was almost midnight before I finished baiting up and we went in for a fine dinner of poached king salmon. Just put a half-inch of water in a pot, and when it's boiling put the salmon steak in for a few minutes; serve it with wine and lemon sauce. All winter I look forward to the first fish of the season. You'll never get a fish in a restaurant as good as we get here three and four times a week.

*MAY 22*—Continued fair. Nine fish by 11 a.m., but then the fishing went dead, so we headed into Cyrus Cove for a picnic lunch ashore. There was logging here years ago, and the shore is strewn with rusty cables and equipment. Fished for a handful east of Eagle Island in the afternoon, then ran back to Camp Island at dusk with a fine brilliant sky to the west.

*MAY 23*—The day came foggy, cool and still, but the fog burned off to a bright day with glassy water. Not much doing, I changed flashers from chrome to white, but there was little difference. We hardly ever get much these bright still days. Gave it up early and ran with Bruce to Sarkar Cove; the town of Deweyville shows on the charts there. The entrance to the cove is narrow, twisting between the steep hills; the cove is like a lake in the woods. Of the town only a single swayback cabin remains on the beach. Piled into Bruce's skiff for the trip ashore, with Sam swimming alongside, trying to get in. Had to beat him off with an oar. Close call.

Through the evening we walked beside a rushing creek. Sat on a log for a long while with the rush of water in our ears. Across the creek two mink played, oblivious of us. Walked back through the forest, which

*Above—The remains of Deweyville, the site of an old wharf and cannery in Sarkar Cove, Sea Otter Sound. Opposite—Kestrel trolling in Sea Otter Sound.*

68

Doreen *and* Kestrel *in Sarkar Cove.*

was like a cathedral with the last rays of the sun shafting down through the tall cedar and spruce. Came out of the woods to a purple dusk, and I sat on the beach as Bruce rowed the girls back to the boat, his wake rippling across the still waters. So peaceful, so peaceful.

*MAY 24*—Day came with a squally southwester and the smell of nearby rain. The *Cape Hason* called on the radio at 6 a.m., and Bob said it'd be worth our while to join him, so we ran all the way to Gas Rock to troll in the corner there. The three of us—Bob, Bruce and I—circled the spot, getting a fish or a hit on almost every other pass, and there was not another boat in sight. As I write this, we lie at the float at Tokeen, a disused cold-storage plant in a forgotten side channel. There's a fish buyer here with a little store and even a shower, so all of our needs are met, and we're hardly two hours' run from the fishing grounds. Damn, we're all lucky to be in here, catching a few fish on the inside and having the time to stop every now and then, taking an afternoon off like yesterday. Can't remember ever having a better time fishing than these past few days.

*MAY 25*—The weather clearing with blustery airs southwest. Ran through the dawn this morning; cast off from the dark, still float and ran with the mist-shrouded islands on all sides. Then the sound opened up just as the yellow sun jumped over the horizon. If we live for special moments, then surely that was one. But it was spotty going today after yesterday's good showing—fished all day for just a handful. Dropped the pick at the head of Holbrook Arm tonight; hills very steep on all sides, and the roar of a waterfall close at hand. Pilings on the beach are all that's left of some forgotten settlement. The *Kestrel's* back at Tokeen tonight, and the *Cape Hason* is holed up somewhere else, but we compared scores tonight on the radio after dinner.

68

*Chuck's admirable fish is a
58-pound red king salmon
taken in Sea Otter Sound.
That's $80 hanging there.*

*MAY 26*—Worked the 26-fathom edge from Holbrook Arm to Edna Bay this drizzly gray day. Smooth bottom on this side with just a few pinnacles to watch out for. Nothing really steep, though, so just sped the boat up and trailed the gear—flew it—over the tops of the underwater peaks. Eight fish by 2 p.m., then nothing for three hours, so gave it up early for a long walk ashore at Marble Island. There's a quarry there, and years ago ships went there to load huge marble blocks for buildings in Seattle and other cities. Now the woods have taken over, and all that's left is rusting machinery, stone blocks scattered here and there and roads overgrown with chest-high Queen Anne's lace. Glad to quit that place and get back to the boat, because the quarry pits are deep, and covered right up to the edge with vegetation. I worried about Sam charging through the woods and maybe falling into a quarry, and we'd have been helpless to do anything as he struggled in the deep green water many feet below. At low tide this evening I dug a bucket of clams in just a few minutes for a fine chowder with dinner.

*MAY 27*—A peach of a day with a clear blue sky and light westerly airs. Worked back to Camp Island and dragged for a fine bite in the afternoon, but the fishing went flat at 6 so we tied to the log boom early for an evening of socializing with the *Kestrel* and the *Cape Hason,* with perhaps not another soul within 20 miles. A little sun and a few fish and we all feel on top of the world. Sometimes it seems as if we give up a lot to drop everything and go north every spring, but on a day like this it is worth it all and more.

*MAY 28*—Fair and hot again, but our little world was shattered today, and just when we had everything, too. We laid at Tokeen in the lazy hot afternoon, having sold our fish after a couple of good days; we were sitting in the sun drinking beer and feeling pretty pleased with ourselves. I walked up to the store with Susanna, and when we returned, everything was in a turmoil. Bob's little daughter, Jacie, five years old, had fallen headfirst into the fish hold, and lay pale and sick. She had all the signs of a concussion, her right arm and side seemingly paralyzed. Bob got on the radio at once, and within an hour a big floatplane splashed down into the cove, picked them all up, and left with a roar that echoed across the bay. It was a pretty morose group that sat around the radio tonight,

Cape Hason (*left*) *and* Kestrel *in Sea Otter Sound, "with perhaps not another soul within 20 miles. A little sun and a few fish and we all feel on top of the world."*

waiting for some word. That little girl was the apple of all our eyes. The news came at 10: the southbound jet was making an unscheduled stop to take them all to Seattle; her condition was serious. Bob asked me to run his boat up to Point Baker, because he didn't know when he would be back.

*MAY 29*—The day came gray, and all of us were about as low as the clouds that pressed down on the water. I got under way at 5 a.m. in the *Cape Hason,* the *Kestrel* a quarter-mile ahead. Made Point Baker at 3. Tied up the *Cape Hason,* picked up our mail, and got in with Bruce and Kathy for the long trip back down the gray strait. Tied to the float again at Tokeen a little after midnight after a tiring and discouraging day.

*MAY 30*—Up and running at 5, past rafts of sea birds rising from the water, past islands and through narrow channels with the rosy dawn all around, with not a boat, not a cabin, not a sign of man in sight. Tried to get Susanna up to see the glories of the day but no luck.

Fished for 10 hours, in the deeps and in the shallows, tried about every kind of bait and gear that I had, but all for nothing. First skunk day in over a year. All day a trickle of boats passed through the sound from Noyes Island and other districts to the south. One was a neighbor from Port Protection; he said boats at Noyes were having two and three skunk days in a row. Tied up early at Camp Island where I lost three straight hands of cards to round out the day.

*MAY 31*—The dawn was fine—yellow and gray above the glassy water—but the fishing was poor again. We had one fish at the tide change, and the *Kestrel* doubled us with two. Steady stream of boats now, all talking on the radio and looking for fish. Another day like this and we'll be running, too.

*JUNE 1*—Gave it up today in Sea Otter Sound and ran—out to the Gulf of Alaska and up the coast to where Chatham Strait opened up like a vast canyon, with the snowy wall of Baranof Island to the west. Sad to leave that sound, you bet. We had it good there for a week when others were scratching, but it went flat and we had no choice. So we ran today, after fish that I heard about a week ago and are probably long gone.

*Signs of a more prosperous settlement at Port Alexander, a tiny bay at the southern tip of Baranof Island: the store (left) and the schoolhouse door (below).*

But what a day for it! The sky was cloudless, the water still, the whole coast painted across the horizon. We ran almost eight hours to slip in here at Port Alexander, and I shut everything down in the still evening. There was no noise except the faint rush of a breeze on the mountain and the soft ticking of the hot engine cooling. The settlement is a row of tired buildings on a spit between the harbor and the empty strait to the east. The ocean is close here, the trees twisted and stunted from the winter gales. The one dusty street was empty, but somewhere a radio played, so we followed the sound to an old man who came down after a bit to pump us some fuel. He said there would be no buyers and no ice in the whole of Chatham Strait until June 15.

The place reminded me of another settlement that I stopped at long ago on a hot summer afternoon, where the marshes of Delaware Bay stretched for miles in all directions. I remembered a winding creek, the buzz of mosquitoes, a general store, and, inside, a group of men talking by a faded poster of F.D.R. A woman in a Mother Hubbard had come out into the heat and the glare to pump gas for my boat.

Here, the mountains rose to snow across the narrow harbor, dusk came quickly, and the night was chill.

*JUNE 2*—Under way at 5, the sky pink to the east, each day longer now as the solstice approaches. Trolled slowly north along the bold shore. The *Kestrel* was just ahead and to the east, but except for that we were alone; I could see islands 30 miles away, and not another boat anywhere. At 8 a.m. Port Conclusion opened in the mountains to the west; it's the bay where Captain Vancouver finished his survey in 1794. Here at the southern tip of Baranof Island it's grim, lonely country for the most part, the hills rising thousands of feet from the water to icy peaks. What harbors there are are steep-sided fjords with deep water and a few anchorages.

One little king salmon by 11 was all we got, so we picked up and ran to where the mountains parted and Port Walter yawned away into the interior. At the head a creeklike channel opened up between the hills and, our wakes rippling across still waters, we passed slowly

*Overleaf—An old herring-reduction plant lies asleep at the head of the inlet in Port Walter.*

*We run into the interior basin and land at the abandoned herring plant in Port Walter.*

into the basin beyond, hardly able to hold back our amazement. The basin was perhaps two miles across, and on all sides the walls rose steeply to snow. Trees clung to the rocks here and there, but mostly it was just bare rock walls. In one place at the foot of the hills, a cluster of buildings and tanks stretched across the flats, but the buildings were deserted and empty, with not a sign of smoke from any chimney. We tied to a rotting float and took the afternoon off to walk through the deserted plant. Snow lay deep on the north sides of the buildings—and this the first week of June! Pictures were on the walls, clothes in some of the bureaus, as if the people had just up and left.

Today with the sun bright and warm, this is a spectacular spot and more. But we're deep in the interior of the island, and the outer bay is one of the rainiest on this entire wet coast: 230 inches of rain a year. I wonder what the winters must be like, with the bay frozen over solid, and the sun gone over the mountain from November to March.

Lay off a snow bank in the afternoon and hauled skiff-loads of snow for our few fish. Bruce's knot

slipped on the third trip and we both went chest deep into the icy water to catch the skiff, gasping with the cold and laughing all at once. Went to the government fisheries station tonight, at New Port Walter, to take water. Across the bay, the ruins of yet another cannery. Almost a week now since any real fishing. The night is starry and very hushed.

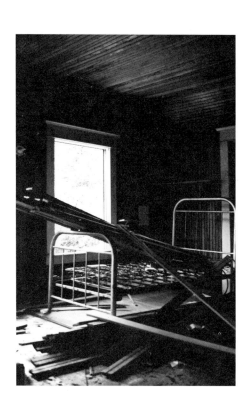

*"Southeastern Alaska
is a hard country
to make a living in,
and the ruins in bay
after bay are constant
reminders of it."*

# The Ruins

When we're running past bay after empty bay, the country here seems lonely and desolate, little settled or visited. But if we go ashore and poke around in the underbrush, we learn a very different story. I've been in hardly a bay that didn't show some sign of previous settlers if I looked around long enough. And for the most part, the stories of these places seem to be all the same: of fishermen and prospectors, of trappers and miners, all trying to make a go of it in an inhospitable land, and almost all failing.

In the last century, and the first part of this one, prospectors combed the region, looking for gold and minerals, even for good marble deposits. When the finds were promising and the price right, they dug a mine or opened a pit. Some mines were small, just a shaft and a shack, and others were whole towns in themselves, with docks and railroads, stores and schools, spread out over the shores and hills of an entire bay. But few

*Right—The workers' cabins
at the Hood Bay cannery.
Below—A herring plant at
Bay of Pillars.*

lasted more than 10 or 20 years, and most less. The deposits were thin, the market collapsed, or the prospectors' luck ran out. When it did, the towns died. There was nothing to keep the men, and greener pastures beckoned.

Usually a few families stayed on, to try and make a go of it, maybe living off the land or caretaking, perhaps all alone, as the settlements turned to ruins around them. But in the end the isolation and the long dreary winters got the best of most of the families who remained, and the rain and the forests soon took over what was left of the settlements. Now only the largest of the abandoned communities show anything from the water: perhaps an old steam engine rusting on the beach or a few rotting pilings sticking up from the water.

The mink and fox farms went pretty much the same way. With fish in the streams for feed, an island in Southeastern Alaska was an almost perfect place to raise mink and fox if the price for fur was right. And if you didn't mind the isolation, a man with a family could have a pretty good life. But then the fur prices fell, the government said it was illegal to catch the fish for feed, and pretty soon the farms were all abandoned, the families moving on to something else. Many's the island where I've gone ashore and walked through gardens and houses, past rows of cages and sheds, silent tributes to some family's years of work and patience. Now only ruined buildings with an air of sadness look out on empty bays.

Most of the larger ruins are not farms, but fish plants of one sort or another: salteries, canneries, whaling stations, herring plants. To travel

*Above—In the house of
an abandoned fur
farm; Harbor Island,
Holkham Bay.
Overleaf—An old
homestead in Masden Cove,
Chilkat Inlet.*

through the bays is to go back through the history of the fisheries in the region. All boomed for a while, and the little settlements dotted the coast. But the fishermen put no fish back and saved little for the future, and pretty soon there was nothing left to catch and the plants were abandoned and the forest took over.

Of all the fisheries that boomed in the past, only salmon fishing has survived in anything like its past glory. And even that is much reduced, with just a few canneries where once there were dozens. Part of this is due to the natural process of economizing and streamlining, brought on by the fast refrigerated packers that can run the fish to a few centrally located canneries. But a larger part of the reduction of the salmon fishery is due to the depletion of the salmon runs. When management was slack and fish plentiful, the big runs were fished until there was nothing left, and then the boats moved on to better areas and only the canneries were left. Just this year, one of the last big outlying canneries, the Chatham cannery, went into caretaker status, and the company started to strip all the equipment out of it. I doubt that it will ever reopen.

Southeastern Alaska is a hard country to make a living in, and the ruins in bay after bay are constant reminders of it.

*Top—The herring plant at Bay of Pillars, going to seed. Above—Bruce and Kathy gaze at a giant piece of floating junk in Chatham Strait. Right—On the move again:* Kestrel *in Keku Strait.*

*JUNE 3*—Off in the chill predawn at 4 a.m. to try the far shore at first light. I can hardly describe slipping out into the still strait this morning, with not a light on any shore and just the palest trace of pink dawn on the snowy mountain wall behind. The day came overcast and chill, the strait empty but for a single tug and tow hull-down on the horizon to the south. Tried Port Malmesbury, from the deeps all the way up to the beach, for nothing. Moved north to Tebenkof Bay for the same results. Finally had two fish in the afternoon in Bay of Pillars, but not much size to them. Bruce doubled me with four fish. Anchored tonight off the old reduction plant at Bay of Pillars. There's a rotting dock here and the usual buildings and rusting machinery on the beach, all somehow depressing in this setting of lead-gray sky and water. Just to the north in Washington Bay is another abandoned plant. Once, there were herring in great schools in Chatham Strait, and the salmon—kings and silvers—fed on them. Port Alexander was a big town then, a center for the

trolling fleet that fished the big runs that came around Cape Ommaney, at the southern tip of Baranof Island. Then herring plants were built up and down the strait and seiners came up from California to fish. In little more than a decade, the plants were closed; the herring were fished out and the salmon gone, too.

*JUNE 4*—The day came cloudless, still. To pull the anchor in the pale predawn off another lifeless cannery, slip out into the empty strait, put the gear in the water before the sun comes over the mountain, just our two boats in maybe a hundred miles of shore-line—this is what we came for, so we savored the morning and enjoyed it.

Bay of Pillars, Washington Bay, Kingsmill Point, Security Bay: we tried them all for just a few little fish. Bruce and I pulled our gear by Cornwallis Point and lay bow to bow and talked for a bit in the hot and still afternoon. Our Chatham Strait trolling trip was hardly paying our fuel, we were low on bait, and there wasn't a buyer for miles. Susanna had been itchy to get

Right—Jacie Dolan, back
from surgery in Seattle for a
fractured skull. She's O.K.
Below—From Kake, a
30-mile view across Frederick
Sound and Chatham Strait
to Baranof Island.
Opposite—A glasslike
Sumner Strait near Hole
in the Wall.

back to the cabin and all of a sudden I was, too. So we ran a few lazy miles to Kake, an Indian village where we lie now on a warm and dusty evening, with children and dogs playing in the single street, and a brass sun going down in the mountains across the strait. From a tired old towboat I scrounged a chart of the little-used channel that winds through the hills and rockpiles to the south, the shortcut back to Sumner Strait and our cabin. Bruce and Kathy are coming, too, so we'll have the *Kestrel* for company.

*JUNE 5*—Under way at 5 on yet another gem of a morning. Across the channel, looming eerily above the trees and water, pink in the rosy dawn, towered Baranof Island, 30 miles away. Then we were in the pass, weaving in and out of the flats and rocks, in the very shadow of the trees—twice my poles knocked branches down on the deck. There was a warm land smell in the air, the alders and birch bright green on the shore. We even saw Mr. Bear today.

The day was so fair that we stopped at a long-deserted cabin in a small sheltered bight, Sumner Island, for lunch. We lay on a grassy knoll and got pleasingly sunburned for the first time in two years, with the wind in the trees and the cries of the gulls in our ears. When the sun grew cool and dropped in the sky we were on our way to cover the last miles to Point Baker, where we tied up at 9 after five weeks away.

On the flats behind our cabin, the grass was green, the garden had sprouted, and the flowers were all out With the gill-net season opening in this district in two weeks, a float to build and a net to hang, we'll just stay here.

Best of all, Bob and his family were there, his little girl bandaged but well, and we all breathed easier.

So ended our Chatham Strait troll trip. We fished a lot of country; we hardly made a dime. Yet I'd do it again in a minute. That lonely canyon cast a spell on us. We fished for days, went in and out of half a dozen bays, and in all that time hardly saw another boat.

*JUNE 6-10*—These have been beach-logging days at Point Baker. Susanna works in the garden and each morning Bob, Bruce, Sam and I set off down the beach with a boatload of chainsaws, jacks, axes and peaveys, looking for logs to build a float with. The ones we want are huge boomsticks—70 feet long and 3 or 4 feet thick at the butt. Each day we work in a different spot, sawing and jacking, laying out skids and clearing the way to the water. The days have been gray with frequent rain, and the work has been hard and frustrating, requiring a whole day for each log. But it's been good work, and we laugh and joke working those rocky beaches between the dark trees and the empty strait. The best is the end of the day, when the jug comes out for the long trip back to Baker and a huge log trails behind the boat.

*JUNE 11*—Last night we towed the last log around the point at midnight with the tide. Put it all together today, in fair, warm weather. Cut and notched-in the cross logs and tied them all together with heavy cable

*Above—Beach-logging at Point Baker. Right—Moorage at Port Protection.*

and staples. By night we had a raft ponderously moving up and down with the long swell. Bruce and I did it, and when we were done we walked back and forth on it.

*JUNE 12*—The *Kestrel* and the *Cape Hason* were off at first light for three days' trolling. I went down the back channel with Susanna to spend a quiet afternoon at Port Protection, stripping the corkline off last year's net. Hummingbirds darted in the trees and two kingfishers swooped for fry in the bay. We worked until the water was red with the dusk, and headed back in the skiff as the lights came on in the cabins around the cove.

*JUNE 13*—Overcast with southerly airs. Built a shed on the float today with left-over lumber from the cabin. Laid out the net to start in on first thing

tomorrow. Susanna says Sam is better than a rototiller in the garden, and cheaper too. Sat by the fire tonight, looking out at the strait, and at dusk the first cruise ship of the season passed northbound. The deep-water route north takes them past the point, and four and five a week pass in the summer.

*JUNE 14*—The whales came today, just about on schedule, too. Walked through the woods to tell Susanna, and she spent the afternoon watching them from the end of the float. Humpbacks, she said tonight at dinner, a mother and a calf. The strait turns the corner by the cabin, and the strong currents stir feed up to the surface. Every year a pair of whales spends the summer there, playing and feeding in the rip.

Worked on the net all day, with the little radio on as I worked. At times I could hear Bob and Bruce talking to each other 50 miles away. Tried to call them, without luck, but it feels good to be working here and listening

*The summertime view from the cabin into Sumner Strait offers the sight of the cruise ships (here the* West Star) *and the resident humpback whales.*

to them talk. Worked until dark on the net, ate dinner, brought out a lantern and worked some more. Nothing like doing things at the last minute.

*JUNE 15*—Cloudy with rain. Finished the last of the net at 3 p.m., all 1,800 feet of it, and rolled it onto the big reel on the stern of the boat. There's something special about the sheen and smell of a brand-new net before it goes into the water for the first time. The zoo starts tomorrow, I guess; boats have been trickling into Baker for the last two days for the opening of the gill-net season. We read in bed for a bit tonight, and in the stillness, the breathing and splashing of the whales carried all the way to where we were lying.

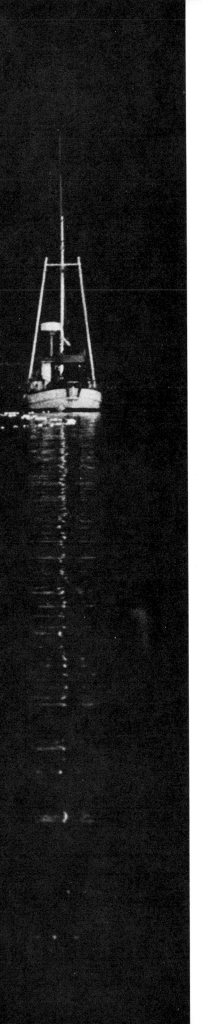

*"It seems a simple
way to fish,
but in reality
it is not. . . ."*

# Gill-Netting

Until the limited-entry law put the lid on new licenses in 1975, gill-netting was the fastest-growing fishery in Southeastern Alaska. The boats are small, generally less than 40 feet, the investment need not be great, and the whole operation can be handled by just one person.

The gear is simply a wall of nylon net with corks on the top and lead-weighted line sewn onto the bottom. The net is stored on a large drum or reel on the stern of the vessel. To set the net, the gill-netter throws the end overboard with a large marker buoy or net light attached, and the boat steams away, the net unrolling off the reel. In the water the net hangs down from the surface like a fence, and drifts with the currents. The thin nylon is designed to be nearly invisible, so that the fish swim into it, are caught by their gills, and trapped. When the catch is to be hauled in, the net is reeled back aboard by the power reel, past the fisherman standing in the stern of the boat and controlling the reel with a foot pedal. When a

*Bruce and Kathy pick the net in the stern of the* Kestrel *and haul in the sockeyes.*

fish comes over the stern rollers, he stops the reel and untangles—or picks—the fish out of the net. On calm days, the boat may stay attached to the net and drift with it. But when there is a breeze, the pull of the boat will distort the shape of the net, and the fisherman should untie the net and idle, or jog, near it. A set may last from a few minutes to a couple of hours, depending on the tide, the location and other conditions.

It seems a simple way to fish, but in reality it is not. The nets are long—1,200 or 1,800 feet—by 30 feet deep, and are vulnerable to damage by small drifting objects (drift), by floating logs, and also by the swift tidal currents themselves. The best fishing is often near the rocky shores, where the tide runs the hardest, and this puts a great premium on timing and experience. A good fisherman might swing his net past some little point in a fast tide and scoop out a hundred fish; a novice might try the same trick and in a minute lose the whole net as it's ripped on the rocks or sucked under by the current.

In the net fisheries, fishing time is restricted in order to allow enough fish to get up the rivers to spawn. A typical gill-net fishing period might be from noon Sunday to noon Tuesday, 48 hours of fishing around the clock. It's not a 9-to-5 job; if there are just 48 hours to make your living, you'll fish 48 hours and rest when it's over.

There are five species of salmon that pass into the waters of Alaska: kings, silvers, dogs, sockeyes and humpies. Each goes—runs—to different streams at different times. So our fishing is always in spurts, as the different runs of fish pass through the area on their way to the spawning streams.

*Above—The gill-netter* Silver Crest *on the hook near Yellow Island.*

In many areas, and especially Sumner Strait, the best fishing is at a certain stage of the tide, which only comes once every 12 hours; furthermore, the best fishing is in only a certain section of the fishing area, and that might vary from tide to tide. The result is often very aggressive and competitive fishing, as many boats crowd into an area with room for a few and jockey to be in the right spot at the right stage of the tide. There are few rules, and during hot fishing, conditions are close to cutthroat, and confrontations frequent.

Still, it can be an enjoyable way to fish, and for most fishermen, running your own boat and being your own boss is hard to beat. It is an excellent fishery for couples, and more and more boats have women on board, in a welcome change from the days when salmon fishing in Alaska meant leaving a wife behind for months at a time.

*Susanna and Sam (upper right standing on top of everything) picking sockeyes. Susanna runs the drum and Sam barks them aboard.*

*J*UNE 16—Checked the oil, greased the reel drive, crossed my fingers, and went out into the strait at 10 a.m. with 40 other boats to look for a set and wait until noon. The government airplane made its pass to start the season, and I set the net out in a long curving line off the stern. The tide had turned and was flooding hard, and I was right on the line with no one in front, just where I wanted to be. So the season started; I was scrambling for a set on the typical chill and overcast Southeastern Alaska day. Had 80 fish on the first one—a good set—moved back and dumped the net in again for another 30. Then the flood tide was over and I moved to the north side to fish the ebb along the beach.

Here in Sumner Strait we pretty much fish the tides. The fish are traveling fish, moving with the incoming tide and headed for the big mainland rivers to the east, so the best sets are right after the tide turns to flood. The fishing area is bounded on the west by an imaginary line between Point Baker and Point Barrie, which is on the other side of the strait. Because the fish come from the west, the line's where you want to be when the tide changes. But it's a tricky and competitive business, with many boats trying for the same spot. So you try to set east of the line on the last of the ebb and hope that you will be carried west to the line and stop there when the tide changes.

Dusk was early, with thick clouds and rain. Moved to the north side of the strait and back to the east to fish the ebb. The fleet out in the strait thinned out as many boats anchored rather than fish the ebb tide in the dark.

*JUNE 17*—Fished the ebb last night for 20. Don't know if it's worth it, though, because of the wear and tear on your nerves, leaving the net and running back and forth with the spotlight, looking for logs and big kelp islands. The day came with patchy fog. At 11 a.m. the tender came alongside to take my fish and talk for a bit before we went out to catch the flood again. Two hundred fish in the first 24 hours; that's pretty good at almost $5 apiece. Trolling pays a lot of our bills, but this gill netting—that's where we make our season. The first day is always a worry, because then I find out whether all of the gear works and the net fishes as it should.

Slower going today, or else I missed the fish. Had a close call tonight with a big kelp island that I never even saw until it was right in the net. Lucky to get out of it without ripping the net. Gave it up to drop the hook at Merrifield Bay at midnight with just over a hundred for the day.

*JUNE 18*—Stumbled out to catch the tide. Had a good set, too, right at the edge of the rip. Standing in the stern, pulling in a few fathoms of net at a time to stay clear of it, I could see the fog slowly burning off to a fine bright morning. Noon came too soon, the period was over, and we ran to the cove to lie with dozens of other boats, drinking beer in the sun and waiting to unload at the tender.

Only two days are all we're allowed this week. Last year it was three days a week; before that, four. Two days a week mean only four tides to make your week in, and that doesn't give you much chance for error. Counted 70 lights out there at dusk, and that bodes ill for the future. That's more than I've ever seen here, and the more boats we have, the less time the government will give us to fish.

Think the best part of the week for Sam and Susanna was when we tied up at the float at Baker, when it was all over. Susanna headed for her garden and Sam just beat feet into the woods.

*JUNE 19*—With steady rain, spent our time on inside projects. The dog was waiting out back all day long for someone to throw something for him.

*JUNE 20*—The day was spent in a wild goose chase with the *Kestrel* to Fontaine Island, two hours away, looking for lumber for the float. There's an old mink farm there, but the lumber was all rotting and useless. The afternoon was warm, and we spent it lazily, beachcombing and just playing in the tall grass, the four of us and Sam, who likes these side trips best of all.

*JUNE 21*—The big event today was setting the second mooring for the float: a 55-gallon drum full of cement and rocks. A big splash is hardly the term for what happened when we pushed the drum over the side. We had hummingbirds at the feeder tonight and the first vegetables from the garden, too.

*JUNE 22*—High clouds from the south and a dropping glass; we expect a breeze tomorrow. Caulked, copper-painted, and launched Susanna's skiff, and a neighbor said he wished his girl friend had as pretty lines as that skiff has. Found Mr. and Mrs. Swallow building their nest in the shed on the float too.

*Susanna's skiff on its mooring below the cabin at "Port Upton."*

*JUNE 23*—The day came gray and cold, with a southeast wind at 20 knots and rain. Dragged the end of the net into the shed to patch last week's hole. The rain slashed on the roof, and the float beneath the shed tugged against its moorings in the tide, but inside, the shed was cozy. I had put in a window that looks out on the bay, and the overhang keeps out the rain. So it was a peaceful day. Twice a skiff passed down the channel toward Port Protection, but other than that there were only the rain on the roof and the occasional call of a raven in the tall woods to keep me company.

*JUNE 24*—Gill-netting; the weather again poor. Started out on line in the usual spot but I junked out—the net fouled with floating debris—in just a few minutes for only a few fish. Picked and moved back, for the same treatment. The strait was covered with kelp and popweed and drift pulled off the beaches and ripped off the rocks by the storm and the big tides. Fished short drifts all day, trying to keep the net away from the worst of it. A fisherman on the radio said he lost half his net in the rip; saved what he could and cut away the rest. Didn't much feel like taking a chance in the black, so I ran in behind Yellow Island at midnight, tired and disgusted. Still, I had 110 for the day.

*JUNE 25*—Up at 4:30 a.m. to catch the end of the flood. Day came with spits of rain and low racing clouds. Junk everywhere; got hit so bad at Point Barrie that we were a mile over the line before we could even get the net back. Susanna and I were in the cockpit, throwing kelp right and left, and Sam was on top of the pilothouse, barking all the while. Bagged it again at dark to slip into Baker, where boats were tied three and four deep at the float and the tiny bar was full.

*JUNE 26*—Woke at 4 to Bob's voice on what we call the sleeper channel, the frequency we leave the radio tuned to at night so we can be reached. The morning was very ugly, but I found a spot on the line and stumbled out to dump my net, still groggy from too much booze last night. I let Susanna sleep these mornings; that way she can watch the net later in the day while I take a nap. That old sun came out just as the fishing period was over, to warm us as we lay in the bay waiting our turn and trading stories. When I'm out there fighting the junk and looking for a set, the whole business gets pretty discouraging sometimes, but at today's prices, even a few fish at a time add up to some kind of a week when it's all done.

*JUNE 27*—We're at the cabin. I slept-in this warm morning, and Susanna and Sam were long gone by the time I made it down the ladder. Spent the day overhauling and rigging the engine for Susanna's skiff. We've got a big, air-cooled, one-lung Briggs engine, a hard-starting beast with direct drive and no reverse. Landing consists of killing the engine about 50 feet from the dock and hoping for the best.

Talked to Flea today, the oldest hand-troller in the bay. He said there's not too much happening out there yet. He goes out every day, though, just like clockwork when the tide changes.

Sometimes we get young guys in here, flat broke and wanting to learn how to hand-troll. It's always Flea who takes them out, shows them the ins and outs of the business, and even gives them his fish money until they get boats and get on their feet.

Our neighbor, Flea, makes his living hand-trolling with his tired old skiff and just one line. The gurdy is plywood and pipe fittings, the gear is controlled with a screwdriver handle. Every day, he heads out into the strait alone on the tide.

*The little settlement at Port Protection is gradually falling back into the forest.*

*JUNE 28*—Had a long visit at Port Protection today, the weather warm and unseasonably fair. The village is about three miles away, through a narrow back channel and across a bay. About 20 people live there, just about like Baker, in cabins set back in the woods around the cove. It must be as pretty a bay as there is on the coast. Beyond these two communities, the nearest town is some 50 miles away.

Stayed till midnight, then headed back across the quiet bay, with the sunset still lingering in the sky under the clouds. Went through the back channel at full speed on the outboard, and picked out the rocks and turns in the dim light all the way. At Baker, the mailboat was spotlighting its way into the narrow entrance. The mailboat had the lumber we had ordered for the deck of the float; the crew lowered the two bundles of lumber into the water. I tied them off to a piling and headed back through the narrow channel to the cabin in the moonless black.

*JUNE 29*—Birds singing, whales jumping, the sun coming out—what a day! Towed the lumber back, then Bob and Bruce came over and we had at it with hammers and spikes and chain saws. Notched the logs, spiked in the joists, and laid the first plank, a clear spruce 2-by-12. The beer came out and in just a few hours the job was done, a deck 12 by 60 feet, and it seemed huge to us. But with a net rack and two or three nets piled here and there, I didn't think it would take too long to fill it. Already the shed is so full of nets and outboards and gear that I can hardly get in, and the shed's not even two weeks old. We passed the jug around and got good and happy for the gill-net opening tomorrow.

*JUNE 30*—The day came rainy and cool for the third gill-net opening. Glad for it, too. Our nets are the thinnest nylon, in special colors, but on a clear day they stand out in the water like a wall, and the catch drastically declines. Got skunked on the first set, right in the middle of the flood, too. Moved to the north for a handful in the corner, and continued very badly all day. Made a set in the dark for 11 sad little fish, so gave it up to run in to Baker at 1 a.m.

*JULY 1*—Rainy. Out at 5 a.m. Made a three-hour set on the line, perfect set and no one in front, for only 30. Sometimes it's easy to believe that someone's cutting those fish off out in the ocean before they get to our shores. The weather was persistently foggy all day, and we had a very close call with a tug and tow in the afternoon. The tug barrelled out of the fog a few hundred yards away and headed right for the middle of the net. Then the tug's pilothouse door flew open, and out ran a man back to the big tow winch on the stern and released the brakes. With the cable smoking off the drum, the tug blew out a puff of black smoke

*Left—Bob and Sam
towing the lumber to plank
the float.
Below—Jonni hauls the
net on* Cape Hason *while
Bob sleeps.*

*The tug* Mary Catherine II *with a load of preassembled buildings.*

and squatted hard over in a turn. When the tug was clear, the fellow hauled in the tow cable and just about jerked that barge sideways to clear me. I called that man up and thanked him, you bet. He could have just gone through us and taken the net with him without stopping. The night was very black and one of our neighbors was not so lucky—I watched speechless as the Canadian cruise ship, *Prince George,* brightly lit and traveling an easy 20 knots, slid past me and cut his net clean in half.

Junked out badly on the last set. Rolled everything on board and dropped the pick; I'll worry about the mess on deck in the morning.

*JULY 2*—Fog thick again. Out in the strait at 5 a.m., using the radar to set the net, with just gulls for company. Once, fishing alone on such a morning, and without a compass in the cockpit, I set net in a complete circle and almost ran over my own net. This morning, the fog lifted and settled, making the landmarks seem strange and unfamiliar. It burned off and I saw the *Kestrel* a few hundred yards away, so went alongside for coffee. Sat on the deck in the sun. Behind us the nets stretched away in two long curves. To the south were a few other boats, but except for them we were pretty much alone. The end of the period was coming up, and in a bit we'd pick and head in to unload, but for an hour we just lay there, feeling the warmth of the sun after a discouraging week.

*JULY 3*—The day was warm and fair, with crows calling in the trees and eagles soaring above the bay, after two dreary, rainy weeks. There's still no sign of any troll fish here, so I spent day on the float, varnishing the mast for the skiff. Found three dead baby swallows on the deck by the shed—a corner of the nest had given away. A fourth baby was still alive and in the nest, so I nailed up a board to keep him in and hoped for the best, but the mother and father were nowhere in sight. The dusk came red with the evening mists rising from the far shore.

*JULY 4*—Woke to the noise of the whales jumping clear out of the water and crashing down again. Some say they're playing when they do that; others say they're shaking the parasites off their backs.

Big day on the float, with baked steelhead, clams from the bay and veggies from the garden. We're almost living off the land. A fine Fourth of July, with many friends here at the float. One of the older fishermen slowly toppled into the bay, taking Bruce's wife, Kathy, with him; we pulled them out, sputtering and laughing. In the evening we went up to the cabin and built a fire, and sat around with the day dying brilliant red and orange across the strait.

But the fishing period for this area has been reduced to just 24 hours this week, and it worries us deeply. Bob and Bruce are off in the morning to run 18 hours to another area where the fishing period is two days. We should go, too, but don't much want to.

It was almost midnight before the party was over, and we all walked down to the point below the cabin where the skiffs were tied up. Just then a whale blew, in the stillness, inside the rock he was, right off the end of the float. I've never seen them in so close. As we watched he blew again, lifted his tail high in the fading light, and sounded.

*JULY 5*—Day came at 6, with the *Kestrel* honking on the float for us to go. We waved them off, told them we'd stay and try it here; then they were off and gone around the point. They'll probably do better up there, but for us to run 18 hours each way to an area I hardly

*Left—Sandra, Irving and Allan Stein. Below—Eric Schugren loading sockeye salmon on the packer* Cypress.
*Overleaf—The* Cypress *is flanked by the gill-netters* Firkin *and* Kay II *unloading sockeye.*

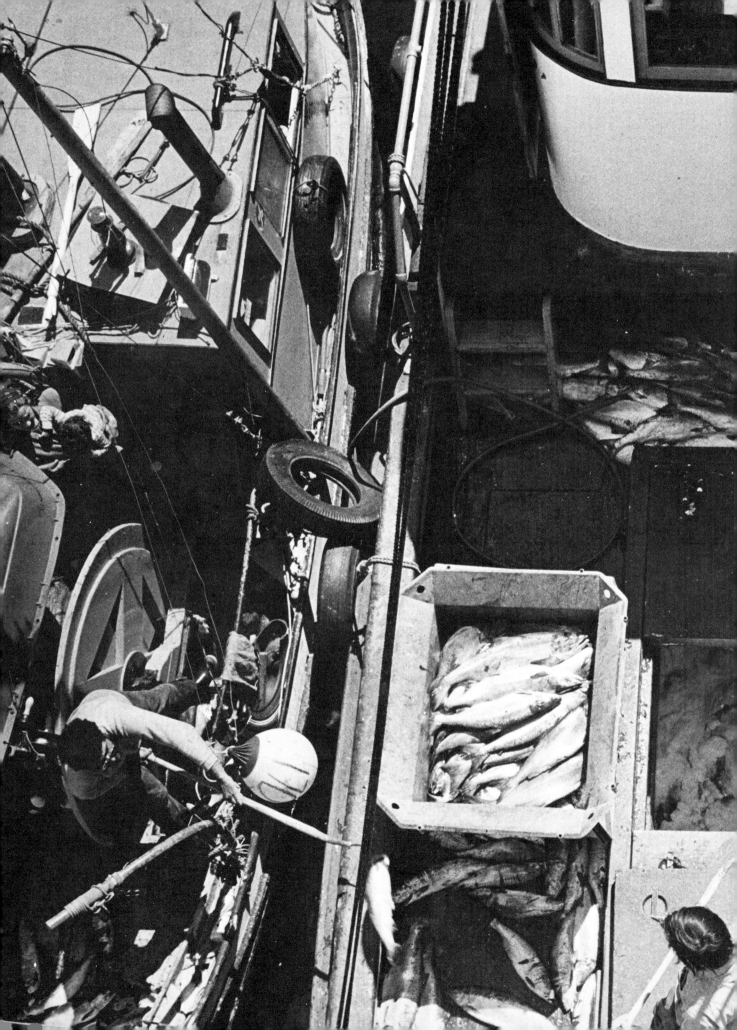

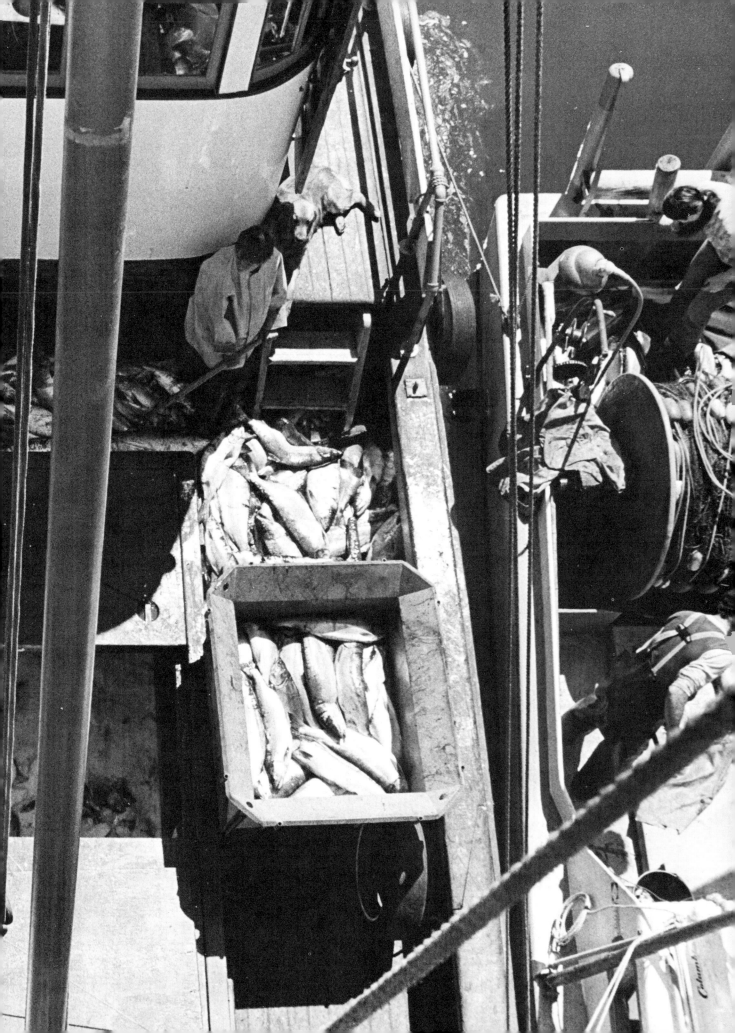

know to fish an extra day seems crazy. Spent a thoughtful day patching the roof of the cabin, the cove and the strait spread out below me as I worked.

*JULY 6*—The day came chilly with a smell of rain in the hills. Overhauled the net this morning and changed a section to a lighter color. I'm very nervous about this 24-hour fishing period; there's too little room for error and not any troll fishing to fall back on yet. The afternoon was clearing and fine after a threatening morning, so we rigged a spritsail to the skiff for a lazy evening sail, ghosting through the back channel and around the cove with the dying evening breeze. All was quiet and still in the cove, just old Flea out for a walk along the shore in the last rays of the sun. Tonight at dusk, as we ate in the cabin, an old double-ender troller slipped into the cove below; his anchor chain rattled, then all was still.

*JULY 7-8*—Gill-netting. Weather fair and warm, the water clear. I could see all the way down to the bottom of the net; the fish could, too, and the fishing was poor. They'd stop and hesitate in front of the net in little schools; we could see them under the boat, and we continually charged the net with the boat to try and drive them in, but with little effect. Yesterday, July 7, had a fine purple dusk that lingered until 1 o'clock in the morning, then the dawn started at 3. On these clear nights in midsummer, the twilight is long, the darkness brief, but after Labor Day each day is noticeably shorter, and by the end of fall fishing, we're using our deck lights at 6 in the evening. Tonight a ghost from the past slipped by, the *Princess Patricia,* another Canadian Pacific cruise ship. Long and narrow, lit up like a city, she slipped through the fleet like a graceful lady, and we all stopped to watch. Huge gibbous moon after midnight, but tedious fishing—I was running back and forth on deck with the light, looking for trash and kelp. Can't afford to miss an hour in these short periods. I'll be glad for noon tomorrow, to sell our few fish, clean up the boat, and take a long nap in the cabin. Now the skiffs are buzzing back and forth in the back channel in the long evening.

*Left—A troller drops anchor in the cove below the cabin at Baker. Opposite—The steamer* Princess Patricia *slips past in the night.*

*"When the fishing period starts, the packers move through the fleet, trying to meet each boat once a day to offload, weigh and grade its fish."*

# The Packers

The fishing areas are frequently far from the canneries, and most fishermen prefer to sell their fish on the fishing grounds rather than ice them and take them to town after the fishing period is over. So, following the boats from place to place, buying the fish and bringing out the mail, are the packers (or tenders).

The packers are generally older boats, 60 feet long or larger, able to pack up to 200,000 pounds of chilled or iced fish. Some started their careers as sardine seiners off the California coast, others as fish-trap tenders in Alaska. A few still have the old slow-speed diesels that most of the fleet abandoned years ago.

A packer's week may start a day or so before the fishing period opens. After buying groceries and odds and ends to sell to the fishing boats, and loading ice, a packer heads for the grounds. Many people on fishing boats seldom go to town; they depend on the packers for groceries, supplies

*The packer* Cypress *flanked by the gill-netters* Firkin *and* Cape Hason *at Merrifield Bay.*

and fuel. There is a lot of loyalty to the packers, and packers develop fleets and follow them as they move up and down the coast with the fish. When the fishing period starts, the packers move through the fleet, trying to meet each boat once a day to offload, weigh and grade its fish. In the evenings the packers lie in some cove near the fishing area, and boats wander in all night long to sell. On a black and windy night when fishing's slow, you'll see packers sometimes with four or five boats alongside or trailing astern, and you'll know the coffee's on and the galley's full.

*Tugs and packers at Petersburg.*

When the fishing period's over, the packers have the long run back to the cannery. In the fall the fleets are way up in the northern districts, sometimes hundreds of miles from the canneries, and they stay there for weeks; only the tenders run back and forth.

When I was a kid I worked on a tender. In the fall we were buying fish up in Icy Strait. We'd finish buying late at night, button everything up and head back to the cannery in Metlakatla, 20 hours away. I'd usually take a watch in the black, and I still remember sitting up there in the

pilothouse, the radio tuned to some Seattle rock station to keep me awake, the green eye of the radar in front of me, and our boat running all alone down some channel with the dark hills slipping away behind.

*JULY 9-12*—For quiet summer days, give me these four. Each day has been like the one before, warm and cloudless. All nature seemed to rejoice in it—sea birds and shore birds soaring and diving, singing and calling, even the whales jumping clear of the water again. Susanna worked on the garden in the morning and spent the warm afternoons with me on the float working on the net. Said the turnips'll be ready in a week if the dog doesn't get them first. The little last swallow finally flew yesterday, tentatively at first, with his parents watching from the mast of the *Doreen,* then farther and farther and finally to the shore, and we felt as if we shared a little in the tiny flier's triumph. Skiffs passed down the channel in the distance, but no one disturbed our little world. Went through the trolling gear, too, to be ready to pick up and run if I have to, for I mistrust this one-day-a-week gill-netting business. While I was working on the boat and the gear, and taking long lazy lunches in the sun, the time seemed to stop. Then the word came over the radio: all net fishing in the region closed until further notice. I had felt that it was coming and had worked to be ready so we could start trolling. We pulled the net onto the float, loaded a few things down the hatch, and went up to the cabin, all set to leave in the morning.

The evening was chilly with a breeze out of the northwest. Built a fire and sat for a long while listening to the fishermen talking on the radio, all worrying about where to go and what to do, and many talking about making the long run to Puget Sound, 800 miles to the south. The sun died cold over the far hills, and out in the strait another cruise ship passed.

So the easy summer fishing and lazy weekends are over, for a while at least. We had it good here while it lasted, but now we feel like running for a while, to try some new places. Last week was too early, but now we're ready.

*JULY 13*—Of all the fine days this summer, this one stands out like a gem. Got up in the cool dawn; Susanna picked a few last things from the garden, and we were off. Took ice at the fish house, then we were set to go, and would have, too, you bet. But there were the *Kestrel* and *Cape Hason;* they'd run in to Baker deep in the night, and Bruce was thinking of making the long run back to Puget Sound for the rest of the season. The day was warm, with a breeze off the land, and it'd probably be months before we were all

together again. So we put off leaving, packed a big lunch, jumped in the skiff, hauled up the sail, and slipped away, leaving Sam at the float. But he jumped in and paddled after us, barking all the while. The wind and tide were fair in the back channel, and we soon left him behind. Port Protection's bay opened up, and we slipped down it in the heat of the afternoon, shirts off to the breeze, and beer cooling in the bilge. There's a cove on the western shore of the bay, and we went ashore, through a trail in the tall woods, and out onto the rocky shore of another hidden inlet. We lay in the sun on a grassy knoll, with the breeze in the trees high above, saying little and just savoring the afternoon. Later we anchored with a rock off the cabin of a friend, and went ashore for dinner with many of the younger fishermen from around the bay. With the fishing uncertain, and boats running in all directions, it was a last dinner, and we all knew it. It was midnight before we left, to slip across the bay with the motor, the water so still it seemed we even weren't moving at all, the purple dusk still lingering on the horizon.

*Dropped the net and gear on the float and went off in the skiff to Port Protection for a holiday with Bruce and Kathy.*

The day cast a spell on us and I think we all felt it. The season's rushing past and soon we'll all be running just to keep up; it felt good to take a day and stop. But tomorrow it all starts again.

*JULY 14*—Left in the dark this morning, ran across the strait and through the dawn in Rocky Pass, with the sun burning off the haze and the trees close around the boat. It's good to run again, to drop the gear in off some deserted shore. Tried the broken bottom near Eliza Harbor on Admiralty Island. To the north, dark hills rose in an unbroken line; to the east, Frederick Sound stretched back into the interior; to the south were Chatham Strait and the ocean. The sky was clear, the water sparkling, and again not another boat in sight. Had a little handful of fish by 3, but we put down the hook in Murder Cove, for sightseeing's part of this trip, too. Another abandoned cannery

*A falling-down cannery in Murder Cove at the southern tip of Admiralty Island.*

sprawled across the shore on a spit between two shallow bays. On the shore, Sam growled at some bear tracks—fresh and deep—brown bear, and not a bit afraid of man. So we stuck to the beach; if a bear charged me there, with 12 shots in my old .38-40 Winchester I might stop him, but in the confines of the woods, we might not be so lucky. Evening is coming cool, the sun over the mountain early, and pink salmon are jumping in the bay.

*JULY 15*—Trolled the rip at Point Gardner this gray still morning. John Muir passed this way one fall afternoon in 1899. He went from Kake across Frederick Sound to Point Gardner—in a canoe, with Indian paddlers. The weather was poor and they had a dirty go getting across that day. In the evening, camped on the beach and resting, the old chief told Muir that he and his men hadn't slept for three days worrying about that crossing.

A dozen fish by noon, all silvers. Susanna pulled a couple while I ate, but they had little size to them, so we picked up when the tide turned and ran across the strait to the dark and forbidding eastern shore of Baranof Island. Another fjordlike inlet opened up between the mountains, and we slipped in to tie up at an empty float in Warm Springs Bay. Seven o'clock and the sun was already over the mountain. We walked up the trail, past sulphurous hot springs welling up steaming from the rocks, and Sam walked past growling, hackles up. The night is still and damp, and the wilderness seems very close.

*JULY 16*—Up at 6 to light rain and clouds hanging over the trees. Just ahead were two huge fish packers that had run in here sometime in the night to sleep for a few hours. This morning we trolled north along the rugged shore, which was half hidden by mists and fog. There was a mysterious quality to it. Now and again the mists would lift to show dark steep hills, with alders here and there along the beach, and the creeks tumbling down, still full of snow melt in July.

So for a peaceful morning. Susanna fished for a bit while I brushed the tangles out of Sam's coat. A few fish around, enough to pay our expenses, but not enough to keep us, so we trolled north steadily all afternoon, circling back once or twice, and following the depth into inlets that opened up to the west, and then out again. Crossed the strait and into Hood Bay at 6, past the huge buoys left over from the days when

*Still, but alive, is Warm Springs Bay on Baranof Island. Gone, the manager's office at Hood Bay cannery.*

deep-draft freighters traveled in to the cannery at the head. But all is silent now and deserted. We went ashore to walk through yet another tumble-down cannery, to sit for a bit in the old store, a breeze rustling the receipts and bills scattered across the floor. The sun broke through the clouds and streamed through the windows for a moment before it dropped behind the mountains, leaving the evening purple and still before us.

*JULY 17*—Up with the dawn to troll slowly out of the bay, drinking coffee in the stern, watching my lines and the dark shores as they slipped past. The weather and fishing were like yesterday, but this time we followed the Admiralty Island shore. Had a lunch stop at Killisnoo Island, to walk past empty buildings and yards grown waist-high with cow parsnip. We enjoy these stops, going ashore and poking around these old canneries and settlements, but three in two days is depressing. Dragged 22 fathoms in the afternoon on the outside of the island for a dozen little silvers. At 7 we passed from the wide strait through the narrow entrance of Kootznahoo Inlet, where we lie

*In Kootznahoo Inlet, the town of Angoon at its head, we find (from left) a home-grown shipyard on a gently sloping beach, a whale blowing in the narrows, and the sad sight of a fish boat abandoned.*

now at the float of the Indian town of Angoon. Mitchell Bay, Kanalku Bay and Kootznahoo Inlet stretch for some 15 miles into the interior of the island, and the three make up a maze of intricate inlets, waterways, channels and islands, some too narrow even for skiffs. The rush of the tides in those narrow channels is very loud, and even as we lie here now, it is a far-away roar.

*JULY 18*—The day came spectacularly fair in this narrow green spot that has houses here and there on the beach. Passed the morning in odds and ends of boat work, greasing the anchor winch and tying up a few shorter leaders to try tomorrow. Then off for a fine lazy afternoon with Susanna in the skiff nosing through those narrow channels, Sam up on the bow sniffing the air. The chart doesn't even show half of what is here—tiny channels with the tide rushing between the trees, down and over rapids to tidal lakes that we could see only faintly. We lunched on a sandy spit, throwing sticks for Sam into the rapids below. Each time he entered, the current swept him off his feet, and he was downstream before he could regain the shore to run back and drop the stick at our feet and wait, barking for it to start all over again.

We putted an easy mile to town in the evening for a walk. The population seemed to be one-half kids, one-quarter dogs, mostly small, and the rest eyes peering out of the windows at us. Sam strutted about like a king. We headed back as the last rays of the sun slanted across the water. A lone whale blew at the buoy before heading out toward the wide strait with the tide.

*Above—Crew members stack the net aboard the Indian purse seiner* Jerilyn *in Chatham Strait. Right—This is the view seen forward from the captain's chair of a gill-netter.*

*JULY 19-21*—Trolling, upper Chatham Strait. The days have been gray, the fishing slow. In the evenings we've been anchored in little bights in the shore. When I got up each morning to pull the anchor, there was always a mountain steep behind me and the strait vast and misty ahead. Trolled north to Point Hepburn, then turned around, suspecting news of a gill-net opening at any time. When the news came, it was hardly encouraging: just 24 hours to fish, starting day after tomorrow. It's a long way to go, but there's nothing to keep us here, so I guess we'll run.

*JULY 22*—The day came overcast, southeast wind at 10 knots; many miles to go, and a poor report for the afternoon. Passed a dozen seiners this morning, making haul after haul off the points, but getting little. The weather was deteriorating, so we lay for a bit in the lee of Point Gardner to pump the skiff and put the poles down, the sky dark to the south, the wind an easy southeast at 25 knots. Ran across in the trough of an ugly chop for three long hours, the skiff taking water

all the time. Halfway across, the skiff's stern was going under, and I knew it was stop or lose it. I spent a wet half-hour with the hand pump, while Susanna jogged the *Doreen* into the steep seas. The tide turned and we had an anxious hour before making the lee of the Keku Islets and narrow Rocky Pass beyond. The sky clouded, and the night fell quickly and caught us in the middle of the pass, spotlighting its unlighted reaches, idling through the kelp in the worst spots. Finally we dropped the hook just north of Devils Elbow, the trolling poles almost in the trees, after a long and tiring day. The jug came out as soon as the anchor was down. Don't know which was worse—crossing from Point Gardner, or feeling my way through the pass in the black with a dropping tide.

*"The handsome and wide-beamed purse seiners, 50 to 60 feet long, many of them series-built and almost identical, are the most distinctive segment of the fishing fleet."*

# Purse Seining

Purse seining, which once accounted for most of the salmon landed in Southeastern Alaska, is now losing its lead to gill-netting. But the handsome and wide-beamed purse seiners, 50 to 60 feet long, many of them series-built and almost identical, are the most distinctive segment of the fishing fleet.

Years ago, seining was an exhausting job on deck, because it involved pulling the heavy net over the rail by hand, and a tedious job ashore, because it meant mending and preserving the cotton and linen nets. Two developments largely changed all that: the hydraulic power block, and the nylon net. The power block is the large pulley on the boom of a seiner. It pulls the net out of the water and lowers it onto the deck, where the crew merely pile it. Nylon nets, stronger and rotfree, last much longer than the old nets. So technology has taken the burden out of purse seining, but it has also made it a more expensive business.

*At the two ends of a purse seine net are a seine skiff ( right) and a big purse seine boat ( opposite). Both are seen in Chatham Strait.*

Purse seining is an efficient and sophisticated operation today, involving long and deep nets and requiring a five- or six-man crew and a powerful skiff. If the fish are schooled (gathered in groups), the net is set around them like a fence. A drawstring (or purseline) around the bottom of the net is pulled (or pursed), drawing the bottom together, pursing the net, and, in effect, making a basket out of it. One end of the net is then taken aboard, and slowly the basket is made smaller until the fish are alongside, close to the surface where they can be scooped out. If the fish aren't schooled, often the net will be set off two points in the water, the net forming a hook shape between the seiner and a power skiff, or perhaps between the seiner and a tree on shore. The tide carries the fish into the bulge of the net and, after a while, the net is pursed. A set might result in nothing, or tons of jellyfish, or thousands of salmon. When fishing is good, it's set after set, from dawn to dark.

The districts open to seining are large, and in the course of a seine opening—fishing period—a seiner may travel many miles, cruising the beach for hours, and looking in bay after bay for a good sign of fish. Some skippers may charter a plane the day before the opening, and fly over the whole area, close to the water and looking for the dark schools of fish. Then the season will start and a boat might end up taking turns setting off a point with a dozen other boats, or in some secluded bay without another boat in sight.

Fishing a few days a week, spending time in town for the rest, often a crew will stick together for the whole season, and maybe for year after year on a good boat. Many men have seined all their lives, perhaps learned from their fathers, and wouldn't think of doing anything else.

*JULY 23*—The tide book says a minus tide for today, and sure enough, woke up this morning and we were anchored in a lake, almost completely surrounded by mudflats. Miles to go and much to do, but no water in the pass, so I had a long breakfast while Susanna rowed ashore with Sam, who jumped over the side to go right up to his chest in the mud. Floated across the shallows at 10 and in to Devils Elbow, where the rising tide was pouring in from the south in a rush. That was the worst spot—the channel's barely 50 feet wide between rock ledges, and there are two right-angle turns within a hundred yards. The water boiled past us, the engine screamed at 2,500 turns, all it had, and we inched through, the boat shearing back and forth in the current. I looked back when we were finally through, and it looked like a river rushing down some rapids.

Late for the gill-net opening this afternoon—just had time to roll the net on, and get out there and scramble for a set in the wind and rain, a long swell driving up from the ocean for the first time this year. Had two good sets, then nothing. Just at dusk I looked to the northwest, back toward the mouth of the pass. It was just one dark unbroken line of hills, no sign at all of the narrow channel we had come through. Night came very black with more wind and rain. At 11 we were clobbered by a mess of kelp and logs that I never even saw until it was in the net, so we slipped in behind Yellow Island to lie with a dozen other boats, and we're heeling over as gusts sweep the anchorage. Those were peaceful days we had trolling; if there were a few more fish, we'd still be up there. This gill-netting is getting to be too crazy for me with these 24-hour periods.

*JULY 24*—Rain again and few fish. Glad to quit at noon, sell, clean up, and head back to our cabin in the drizzly dusk. Did O.K. for the period, I guess, but we had all those fish in only two sets, and it could just as easily have gone the other way. Hope the trolling picks up or it might be a skinny winter.

*JULY 25*—Visited this morning with Byron, a neighbor who's a young skiff troller with a sturdy 16-footer that he built himself. Said he's glad for the summer, when the fish are right out front and he can be home every night. Don't blame him much; in the spring he travels after the fish in his open skiff and camps on the beach at night, and that gets to be a

pretty wet business after a while. The evening came with lifting skies and whales moving inside the rock after dark, blowing and splashing, very loud.

*JULY 26*—Under way at first light to Hole in the Wall. Ahead and astern, half hidden in the mists, two local boats steamed south with us. Hardly an hour's run from the bay, Hole in the Wall is a good spot this time of year, and today half the fleet was there, all the way down to the one-line skiffs, taking their turns on the inside drag. The shore's steep—we dragged in 20 fathoms—but the tips of our poles were almost in the trees. There's an up-and-down bottom here, covered with lines and lures from over the years, but trial and error have shown us where to bounce and where not to, so we had a good day with a few fish, and we lost no gear, either. The fishing tapered off at the turn of the tide, and most of the boats trickled away to the north again. But with ice left over from last week, we stayed until dusk and slipped through the canyonlike entrance to Hole in the Wall. Of all the places we visit each year, this is one of my favorites. It's like a lake in the woods, with grassy flats and steep hills above. The dusk came with muted colors, and a deer slipped out of the woods to drink in a stream. Susanna baked a little king salmon for dinner and we had a quiet evening. The cry of a bird, the murmur of a creek, but otherwise all was still. We did the dishes and sat up for a bit. I don't suppose this bay sees 20 visitors a year. But within a few years the loggers will be here; they're just over the mountains now. And when they come, I don't believe I'll be back.

*Left—Mudflats in
Keku Strait.
Below—Tied to the float at
Point Baker are the skiffs
with which Barry and Byron
hand-troll for salmon under
power of sail.*

*JULY 27*—Weather day—stayed at Hole in the Wall. Woke at 2 to slashing rain with strong gusts williwawing down from the hills. Thought it was a squall, but the day came windy from the southwest with a cold rain, low-flying clouds and a heavy swell breaking across the entrance, unusual weather for midsummer. Trapped, but it's hard to imagine a better spot or company for it. Spent the day in cards and letter writing. The night again was dark and windy, and I lay awake for a while listening to it and to a California station fading in and out on the radio.

*JULY 28*—Day came fresh and clear, with the sweet land smell in the air and the storm blown away in the night. Went out through the left-over swell in the entrance, and I put down the gear just as the sun came over the mountain. Susanna cooked up a bunch of hotcakes and I steered from the stern, drinking coffee and watching the poles for a wiggle. Just six of us here this morning, and it was slow going—almost gave it up at noon and ran—but just before the top of the tide they started to hit, silvers and big ones, 10 and 12 pounds, and they continued into the first of the ebb. Stayed into the next flood, and fished until the flasher started up out on the reef at dusk, when we headed north with 37 for the day.

*JULY 29*—Fair, with northerly airs. Rolled Susanna and Sam out early and sent them through the back channel to Baker with a skiffload of fish to sell and a list of grub to buy. Rolled on the net, stowed the trolling gear, and got off at 10 just as the Port

Kay II *trolls for salmon in the waters of Sumner Strait near Hole in the Wall. Some sun and a light breeze make it a good day for work.*

Protection fleet headed around the point for a two-day gill-net opening ahead.

Fished in the corner today with a hungry crowd of boats in that tight spot. The tide runs hard there, and timing is pretty important. I was too early on the first one; drifted over the line and had to pick. Someone got the spot on the set after that, and the next one too, so it was 8 p.m. before I finally made it out to the set that I wanted, the evening westerly throwing spray clear over the pilothouse as I ran for it. Picked it for 80, mostly big chum salmon. Often at this time of year there will be a run of these fish along the beach on the north side of the strait. The fish'll hug the beach, and there's room for only a few boats. But if you can find the fish and stay on them, you can put in a big week when many of the boats on the south side are just scratching.

The sun went down and the moon came up in a clear yellow sky with a red band over the mountains. The wind was chilly after dark, and high above, the first northern lights of the season did their ghostly dance. It's still July, but there's a real snappy chill to the air tonight. Another month and we'll all be headed for the northern districts to follow the fish up into the cold windy inlets until the fall gales drive us south.

*JULY 30*—Day came clear after a tiresome night of fishing in the black, running up and down the strait looking for junk. Battled in the corner again for another good set, then moved way back to the east for the ebb. The afternoon westerly was right on schedule, but it gave way to an unusually fine evening, with that old hunter's moon rising in the east. It's eerie to run in the moonlight on a still night. The steady beat of the engine and the slight vibration of the prop are there, but the water seems still and motionless, as if something's wrong and we're not moving at all.

How black it is when the moon goes down! Tonight there was a haze over the stars, and the black was thick, not a light anywhere except for the boats out in the strait, and no horizon to be seen. Fished the beach with the radar—ran in as close as I dared and then dropped the net light and the end of the net and steamed away, reeling off the net as I went. Behind, all is the blackest black, the net light a dot in the featureless void.

*JULY 31*—Pale and fair dawn. Napped too long and woke with a start to see the net full of kelp and headed for the beach. Yelled for Susanna, and we jumped in the stern and picked like madmen, kelp and fish flying. Rolled the last 40 fathoms onto the reel, fish, kelp and all, before the swift tide could put us on the beach. Made a quick set on the line, and then it was noon, and I idled across the strait, while Susanna and Sam cleaned up the stern. We did well this week; guess things are on the up and up—a few fish and a little good weather, what a change they make in our lives.

*AUGUST 1*—Back at the cabin. I worked on the float this afternoon, doing the week's mending on the net. The season's barely half over and already our net looks like a tattered rag. Years ago, fishermen used heavier nets, and patched them to use for season after season. Now we use extremely light nylon, for nylon nets fish the best, and then we throw them away at the end of the season.

Bob and his family were here, working on their net, too. They saved a steelhead from the week and we grilled it up on a rusty hibachi as the sun died in a blaze of color across the mountains. So ends the first day of August.

*AUGUST 2*—Another cloudless and fair day. Trolled for silvers with 20 other boats in the rip off Protection Head. Had 14 fine-looking fish this morning, but nothing after the tide change, so gave it up for a long walk with Susanna. Around our little cove we walked, to sit on the point and throw sticks for Sam. A single whale worked the rip out beyond us, surfacing to breathe three or four times, then, slowly, ponderously, lifting his tail and slipping down to the depths. Rafts of gulls and phalaropes, and here and there a few cormorants, moved back and forth in the eddies. The sun did its brilliant thing again, and we walked slowly back for a dinner by lamplight, the sky outside red, yellow and, high above, even green.

The weather continues fair. The old-timers marvel at it, and we all enjoy it while we can. Last year, I could count the warm clear days on the fingers of one hand, and now by my count we have had over 20.

*AUGUST 3*—Just Sam and me today, trolling for silvers in the rips, in and out in great circles, miles across. It's pretty much deep water out here, and I spent the day in the cockpit, trying bait, changing spoons, trying to find the right combination. Our neighbors were all out and we waved as we crossed, but no one really was doing much. We fished mainly at the corner, where the strait makes the turn from west to southwest and leads out to the ocean. To the south, Port Protection opened back up into the hills, and Mount Calder, like a child's drawing of the Matterhorn, rose to snow above the trees. To the east, I could see all the way to the Canadian Coast Mountains, 70 miles away. This is our world, and how it satisfies, at

the end of such a day, to sell a few fish and to sit at the cabin watching the evening sky and the water. Homemade pie and rolls tonight; even Sam had some of each on his plate.

*AUGUST 4*—Big halibut score today—neighbor Allan, from Port Protection, roared into the cove to clean a load of halibut so no one else would see what he had and set on his spot. Swore me to secrecy and gave me a big halibut steak to seal the deal.

*At evening, the boats are moored at the float below the cabin at Point Baker. Sumner Strait is calm.*

*"They are bottom feeders, and are caught on what is known as longline gear . . . set on the bottom and buoyed and anchored at either end."*

# Halibut Fishing

Halibut are large flat, flounderlike fish, occasionally reaching 600 pounds and more. They are bottom feeders, and are caught on what is known as longline gear, although they will occasionally be caught by trolling or hand-jigging gear.

Longline gear consists of long lengths of line, usually quarter-inch nylon, set on the bottom and buoyed and anchored at either end. The buoys are often flagged for better visibility. At intervals in the line are leaders and hooks, baited with pieces of herring, octopus or even salmon. One unit—or skate—of gear consists typically of 1,800 feet of line with a hundred hooks. A skate may be set singly, but it is more common for several skates to be tied together to make up a mile or more of line.

Traditionally, much of the halibut has been caught by distinctive white schooners, 50 to 80 feet long. These boats once ranged as far as the Bering Sea on trips of up to 30 days, fishing 50 to 60 skates of gear a day. The

Halibut boats carry on their sterns the equipment for handling longline gear: a line chute where the gill-netter's reel and fairlead rollers would be. These boats, like the gill-netters, are also rigged for trolling.

boats carry large crews. A declining catch and higher operating costs have forced many of the schooners into retirement, but the higher fish prices have made it attractive for smaller boats to enter the fishery, and these concentrate on the calmer inside waters. Longline gear can be pulled by hand from open skiffs.

At Point Baker, halibut fishing is a sensible alternative or addition to salmon fishing for part of the season. However, the tides there are strong, sucking buoys and flags under the water and making it possible to set and pick the longline gear for only a few short periods each day. But at prices approaching $1 a pound, even a few fish can make a good day.

*Allan's "big halibut score" is a day's pay. Halibut run 50 to 100 pounds for males, females five times that, and the price is close to a dollar a pound. A good day.*

AUGUST 5—Gill-netting. Just about wrapped my net around Point Barrie on the first set. The tide was ebbing hard and I knew I'd have to haul the net fast. Then something went wrong with the high speed in the reel drive, and I had only half the net rolled on when I had to quit and tow full speed to pull the net clear of the rocks as the tide swept us around the point. Diagnosed the trouble and radioed town for new parts to be flown out. Plane came at 6 p.m. with the wrong parts, naturally. Then the clutch quit completely at 10 in the evening, when it was very windy, and we spent three hours pulling the net in by hand. Tied to the float at 2 a.m., tired and disgusted. This was the first breakdown of the season that cost us any fishing time.

AUGUST 6—All the cove was still this rainy morning, with wood smoke rising straight up into the overcast, and a lone blue heron walking quietly in the shallows. In a dark and musty woodshed, under about a foot of rusty outboard parts, I found what I needed to repair the reel drive. It was jury-rigged, but at least working, so we fished the day out, staying in the middle and not taking any chances with the beach. The night was black after a smoky red evening, the strait covered with long seams of kelp and trash, whole uprooted trees washed out of the Stikine River by heavy rains in the interior of British Columbia. Ran in to the anchorage at midnight with four others. Only a few boats were left fishing in the strait; the coves on the far shore were bright cities of anchored boats.

*Above—Flea picks his herring net over the bow of his skiff in early morning sunlight. Left—The tug* Sea Kist *with a tow of logs in early morning fog. Such is the diversity of Southeastern Alaska days.*

*AUGUST 7*—Fished the period out by babying the clutch of the reel drive and working on it after each set. Ended up with about half the catch we should have had. Better than a jab with a sharp stick, I guess. Took the whole pump and clutch apart in the afternoon, and reordered parts from town.

*AUGUST 8*—Woke at 2 a.m. with the sound of engines and a foghorn close at hand. Outside the cabin, the fog swirled thick through the trees, blotting out even the light on the rock. A horn again, closer now, then the glow of a spotlight, and the deep beat of heavy engines idling past the point. Glad to be in here snug, and not out there with a big tide running hard in a black, foggy night.

The airplane came at 3 p.m. and brought the wrong parts again. Different parts, but the wrong ones. So, after two plane loads of wrong parts, I decided to go in and do it right, and we were off at 5, eastbound across still waters, the sun painting the water and the sky behind, and Susanna and I excited at the prospect of town. Night came with pesky thick fog, and caught us right in the worst spot, the 20 miles of Wrangell Narrows, winding between the flats and rock piles to Petersburg. Went through with the radar on and one

finger on the chart, the tide swirling us on, channel markers appearing suddenly out of the fog to rush past and disappear behind. Once, we pulled over to the side to turn around and idle into the current as the big Seattle-Alaska ferry passed, taking up all of the narrow channel and showing row after row of lighted windows, visible just for an instant and then lost in the fog. I confess to having an uneasy moment or two there in the tightest spots, when the markers disappeared off the center of the radar screen, but I couldn't yet pick them out of the gloom ahead. We tied to the float at the machine shop in Petersburg, just at the top of the tide on a black, still night.

*AUGUST 9*—Day came at 7 with a steamer hooting its way into the foggy narrows. Susanna and the dog went off uptown to shop and visit. Petersburg is pretty much a cannery town, neat houses set on the shores of an island with thick woods and water on all sides, but it's the big city for us for much of the year. Replumbed most of hydraulic system today and put in a bigger pump and a new clutch. But I was all done at 6 p.m. and we went down the narrows when the tide served. We headed for Sumner Strait in another purple dusk, and ran the last miles in the black.

*AUGUST 10*—Up at 6 and there's Flea, picking his herring net with the morning sun slanting over the trees. He stopped just long enough to tell us to get our butts out there for the coho bite. So we trolled

Protection Head with many others today, had 42 fish until the fishing went dead at the tide change. Tried to get water at Port Protection tonight—they have the best water around—but there was none at the float; a friend said someone anchored on the hose last week and broke it. Sad business, a settlement like this. Little by little the store and floats are going to pieces here. When we first came, four years ago, the general store was the best in the area; you could get anything there from engine parts to fishing gear. An older fellow built the whole place by trading work for supplies with the fishermen who used to lie in the bay in the wintertime. Then he retired and sold the place, and now the store is closed and rotting, the floats are breaking up, and pretty soon there won't be a place to tie up here.

*AUGUST 11*—Had 35 fish in the morning bite today, mostly on plastics and spoons, but I could see that boats using bait were doing even better. We could have stayed for the evening bite, but each day is shorter now than the day before, and I still had to hang my fall net, so I went in early to the cabin to start on it. Put up a lantern in the shed and worked until the sky was dark and Susanna called me.

*AUGUST 12*—Gill-netting. Hot and fair weather, but a breeze to stir up the water. Started out in the slot between Yellow Island and the shore. There's room for only one net in there, and on the ebb it's a good spot for dog salmon when there isn't much doing anywhere else. Fishing these dog salmon is spooky sometimes, because they hit the net and die without even a

*The herring net provides free, fresh bait for trolling. Neighbor Bob Anderson at Point Baker hauls his herring net and finds it full. The troller* Hawk *baits up to make a run in the fog for salmon in nearby Sumner Strait.*

black the net curved away from the boat, shimmering and glowing. I dropped the hook after midnight, dirty and discouraged, with only 80 sad-looking fish for the day.

*AUGUST 14*—More of the same. I finally quit fighting it and made long sets until the period was over, laying out the net eight miles east of the line and letting the ebbing tide carry me down to the line before picking up. Laid the net on the float in the afternoon and washed it for two hours with a high-pressure water pump before it came clean.

*AUGUST 15*—Checked the herring net this morning and it was plugged with fat bright fish—five big pailfuls in just a tiny rag of a net. We leave it set in the bay for most of the season, and it will go for a week with just a handful of fish, then load up in a few hours

like last night. So I trolled with herring as bait today, for 42 silvers. Fished until dusk, when the fog poured up the strait from the ocean, driving us all to shelter before it.

*AUGUST 16*—Day came with thick fog hanging in the trees around the bay. Off at 6 a.m. with the radar, two other boats following behind. Thick, thick fog all morning, but good fishing, the drag crowded; even the skiffs were out there, calling to me as they passed, asking where the fish were. The fog burned off at 2, and I counted 40 boats out there, from one-liners all the way up to a 50-footer, put out to pasture here—it's too rotten to fish on the outside. While waiting to sell our fish at Baker tonight, Sam found a rotten one to roll in and came back covered with it, wagging his tail and grinning from ear to ear.

*"Hand trolling from fishing skiffs
was an attractive and low-cost
way to enter the fishery. . . . Now
there are quite a few skiffs at the
point, some new, some 40 years old."*

# Skiff Fishing

Point Baker is one of the few places in Southeastern Alaska where a man can earn a living trolling salmon in a skiff or an outboard and be home every night. During August, when the silvers run in from the ocean, the fishing grounds are hardly a stone's throw from the buying scow.

When I first came to Point Baker, there were just two skiff fishermen left from the old days, and they used boats that looked only slightly better than the hulls that lay rotting on the beach. But the old-timers told me that in years past 30 or 40 skiffs were fishing here during silver season; the shores of the bay would be covered with the tents of the fishermen. This was before the development of light, air-cooled gas engines; fishermen then used the old, heavy, make-and-break one-lungers, heavy beasts that would just about break your arm to start. And before that, the fishermen just rowed. I've seen faded photos of the old skiffs with raised oarlocks and a man standing, rowing and trolling. But that all was years

*Fritz Powers and a companion troll off Port Protection in Fritz's 16-foot puddle jumper.*

ago, when you could make a good living at the depressed fish price. Then skiff fishing—puddle jumping, we called it—just about died out in the general decline in the fishery.

Recently there has been something of a comeback of the skiffs, at least at the point. Younger people moved into the community. For them, hand trolling from fishing skiffs was an attractive and low-cost way to enter the fishery. They could drag a hull off the beach, caulk it, paint the bottom with hot roofing tar, patch the worst spots with tin and tacks, borrow an outboard, build a gurdy out of plywood and pipe fittings, scrounge a little line and a few lures, and they were in business. And in not too much time they could be prosperous. When the silvers ran, even a skiff fishing one line might make $100 before noon.

Now there are quite a few skiffs at the point, some new, some 40 years old. In the spring, before the fish run in from the ocean up Sumner Strait and past the point, a few of the men with larger skiffs head down the coast to Noyes Island, to fish and to camp on the beach at night. Years ago there used to be skiff camps up and down the coast; now they're

*Below—On the rail of Byron Easter's skiff is his homemade, and efficient, halibut longline roller.*

starting to come back again. Those are big waters out there, but on a good day I've seen skiffs far out in the ocean, almost out of sight of land. One of the local fishermen told me he took his skiff around Cape Addington once. Said the weather blew up in a hurry, and he took a sea over the stern in the rip, two miles off the cape, and it swamped him. Just an open skiff, with a heavy engine and ballasted with railroad irons. He hadn't seen a boat for two hours, but just when he swamped, a troller appeared out of nowhere, threw him a line, and towed him, keeping him afloat, while he threw over the ballast and bailed for all he was worth.

*Above—The traditional dory, known to fishermen on both coasts.*

If you aren't in a hurry and your needs aren't great, skiff fishing in the summer at the point is a pretty good way to go. You can smoke or can what you can't sell, and when fishing's slow, there's a bar that floats on cedar logs right next to the fish house.

*The net is a mess, full of kelp and drift and tangled from a night in the tide rips.*

*A*UGUST 17—Gear in the water at 6:30 this morning and a fish on almost at once! Fished all day, with Susanna sometimes in the cockpit with me, the sun burning off the fog, and boats all around. How good it felt to fish out there until dusk tonight and then slip into Baker to sell the shiny pile of fish on the hatch.

The silver salmon run is on now, and the harbor's crowded with trollers from all over this part of Alaska; the bar is full until late every night. Some of these boats have come here every year at this time for 10 years or longer, to fish for the silvers right out in front of the harbor, and then to run in here in the evenings to sit around with their friends. In all of Southeastern Alaska, there is probably not another community with the fishing grounds so close. The little store, bar and fish house at Point Baker limp through the winter with the business of the few local boats, but they make their season buying fish and selling supplies in these few weeks.

*AUGUST 18*—After three seasons of close calls, we finally got caught good by the flood-tide rip tonight, and it was all over so fast, we hardly knew what hit us. We had a big set on the district boundary line at dusk for over a hundred silver salmon, and it made me too hungry, I guess. The incoming tide was pouring like a river around the point when we picked up the net, and I knew that rip was out there and headed our way. So I ran a whole hour to the northeast to be clear of it, and set the net. The next thing I knew, the boat was turning circles in a churning mass of water and kelp and logs, and the net was completely sunk out of sight. I thought we had lost the whole ball of wax right there, but we started picking, and the net came up clean, from straight down, the corks squeezed to sausages from the pressure in the depths. The net kept coming, and just when I thought we had it, the last 300 feet popped to the surface in a tangled mass, wrapped around logs longer than the boat, and strung through and through with long streamers of kelp. We picked on that mess for three hours, and I was ready to get the knife and start cutting, but the *Cape Hason* came out of the black to help us, towing the logs away from the net and coming alongside to help us get the rest in. We were lucky; just to the west a boat was calling for aid, half its net lost and the other half wrapped around the boat and its propeller. That was enough for us.

*AUGUST 19*—Woke at 6 to the sound of an outboard close at hand, and looked out to see a skiff cruising the anchorage, two fishery agents pointing at the kelp-covered mess on our stern and shaking their heads. The day came with low clouds rippled on their undersides like sand on a low-tide beach.

Set our net in front of the rip again tonight with the *Cape Hason* and three other boats, leaving the nets and running to the west, spotlighting the dark waters and looking for the tide rip. But we had little ambition after last night; we spooked at the first sign of the rip, picked the nets and ran for the anchorage. The night was chilly; pulled two wool shirts on tonight for the first time this summer. It's only the middle of August, but fall and the cold weeks in the northern districts seem very close tonight.

*AUGUST 20*—Started hanging my fall net this afternoon, a tedious process of sewing the leadline on the bottom and the weedlines to the top of the 1,200-foot net. The waters up north are milky blue from the glaciers, and the fish are larger, so we use nets of a different color and larger mesh.

This morning at the tender most of the boats were getting ready to leave this area and make the long run to Puget Sound, 800 miles to the south, or run north to the glacial inlets of Stephens Passage and Lynn Canal. The fishing dies here at the end of August, but in the northern districts the dog salmon stream in from the ocean for another month. Another week, two at the most, and it'll be time for us to run, too.

*AUGUST 21*—Left Susanna working on one end of the net and Sam tangled in the other end, and I trolled the tide rips for silver salmon again. The best fishing's often right on the edge of the trash-filled rip, and the current's the strongest there. Twice I was sucked into it, tangling my lines in the kelp and sticks, but it was worth it—almost 40 fish for the day. Even found two live herring swimming in the bilge of the skiff this evening, left from picking the net this morning.

*AUGUST 22*—Under way before dawn for Shakan Bay, some 12 miles to the south. Guess we were in the wrong spot yesterday; heard last night that some of the boats at Shakan Bay had 80 for the day. So we

*Overleaf—Gill-netters set out at dusk into the strait near Point Barrie, the southwest tip of Kupreanof Island.*

The trollers Twerp (*left*) and Gruesome *motor slowly off Protection Head, about three miles southwest of Point Baker. Trolling offers a way into the fishery at a cost that is less than some others.*

joined 12 others there and even had three fish on the first pass. It's a rough bottom, with jagged underwater peaks ready to snag your lines and rip them off, but Susanna knows how to steer, and as the boat wove through the unseen rock piles below, I stayed in the stern, pulling and cleaning fish. Had 42 by noon, big, bright, handsome fish, and we fished until dusk for 64 for the day. Anchored up in a little cove just on the edge of the fishing grounds. All around the bay, anchor lights dotted the shores, and above, the hills rose, ridge after misty ridge, back toward the center of the island. Years ago the bay had a cannery, a marble quarry, a fox farm and a year-round post office. They're all gone now, and only ruins are left; for most of the year the bay is empty and silent.

*AUGUST 23*—Trolled for 35 fish this foggy morning. Twice Susanna called back from the pilothouse, and I looked up to see another troller pass close, its hull gleaming through the fog, its radar equipment turning eerily on top of the pilothouse. Radar is expensive, and temperamental, too, but it shows the outlines of the shore on a TV-like screen, and you can fish in the fog when other boats lie in and wait. In a foggy season, you can go a long way toward paying for a radar set with the fishing time you gain, not to mention the peace of mind running in a moonless night.

*AUGUST 24*—Just 3 fish was all, from noon to 3, so gave it up to sightsee in Calder Bay, a little-visited spot with steep hills on three sides. A stream winds through a valley there, and we walked it, across the grassy flats and into the woods beyond. It's a rain forest, the floor a thick mossy carpet; even the stumps are just green bumps after a few years. The stream was full of pink

salmon, or humpies, churning up the shallow reaches and milling around in the deeper pools, waiting for their eggs to mature and their final rush to spawn and die. We sat on a log for a long while, looking at the pool below. There must have been a couple thousand fish down there, swimming a slow circle in one dark mass. There is something mysterious about these dark masses of fish, homing in from the reaches of the North Pacific Ocean to this lonely stream. We watched for more than an hour, until the light in the forest started to grow dim and the steady cold rain started again. On the far side of the flats a single bear disappeared into the woods as we passed, and the night fell quickly, and very black.

AUGUST 25—Gill-netting. Day came with southeast gale warnings for tonight. Talk about problems: last week it was the tide rip, before that it was the algae, this week it's jellyfish in the net, thousands of them, brought to the surface by the rains. The fish came up covered with jellyfish, and splashed stinging nettles in our faces and eyes. One big jellyfish that came over the roller tonight would have filled a 10-gallon bucket. Lowering skies with wind in the afternoon, our world a narrow strip between dark hills and darker water. Fished all night in dirty weather, looked out and counted less than a dozen boats fishing in all of Sumner Strait. All the signs point toward moving north after this one.

AUGUST 26—Rainy. The jellyfish were even worse than yesterday, plugging the scuppers and half filling the cockpit by the end of each set. Had a few fish on the Point Baker side, but we were driven out by the jellies. I've never seen them this bad here.

Fished Point Barrie with the *Kay II* during the dreary gray evening; no other boats were even in sight. With the empty strait fading away into the mists, and the steep hills rising into the clouds behind the boat, our sense of isolation was very strong tonight. Had just three fish for a two-hour drift in the black, so we both gave it up at midnight to anchor, tied together, and start in on a jug of wine, as the tide swished between the hulls in the stillness.

*AUGUST 27*—The period ended, so I picked the last stinger off my face, and we sold our final Sumner Strait fish for the season. Everyone I spoke to at the tender is headed north in a day or two, and we won't be far behind. Susanna dug out the wool clothes, and I worked on hanging the fall net this afternoon, until the dusk was smoky pink. The moon came up over the mountain, and the fog poured up the strait from the ocean again. Today was warm, but fall's here now, no mistake about it. The government might open up more fishing here for another week, but it's time to run, I can feel it. It's sad to leave this spot, the cozy cabin with the fishing area right out the front door. Summer fishing paid our bills, but we make our winter in the cold and windy inlets of the northern districts. A week up there might bring more than two or three weeks here.

*AUGUST 28*—Changed nets today—put away the green summer net and rolled on the ice-blue fall net.

It's new now, but in the heavy fall fishing, we don't have the time to pick carefully, and we rip the meshes as we pick. We might use the net for 10 days at the most, and then it's just a rag ready for the scrap heap. There were odds and ends of work left to do, but the day was fair, maybe our last summer day here, so I took off across the bay with Susanna for a sail to Port Protection where the local boats were finishing up their nets and getting ready, like us, to head north in the morning. A neighbor came out in a seven-foot pram with a plastic sail, and we had the first annual Port Protection-Point Baker scratch race. We'd better enjoy it while we can, I guess; by the time we get back from fall fishing, the weather will have turned and the fall rains started.

*Below—Big Jim Mock, seven-foot pram and plastic sail. It goes good, but Susanna's skiff sails right by him. Right—The end of a day, with a long view to the west from the cabin porch.*

*"After the equinox the
weather changes—violently—
for the worse."*

# Fall

Each year about the third week of August, the summer fishing tapers off in the southern districts—Tree Point, Clarence Strait, Sumner Strait and Frederick Sound. The humpy and sockeye runs have passed, and by late August the catches are dominated by coho and dog (chum) salmon.

Sometimes there is good fishing clear into September, but usually the last week of August means a choice of some kind for the gill-net fleet. Some boats will finish up fishing on a Monday or Tuesday in Alaska, load up their gear and run 800 miles to Puget Sound to catch the opening of the fishing period there at 6 on Sunday evening. Day and night they travel, rushing to Puget Sound to roll on a deep fall net, and many of them make it on time. Others, the older fishermen or those with especially tired boats, stay in the summer districts to fish week after week—each week fewer boats and fewer fish—until finally the southern districts close and they can roll off their nets and call it a season.

*A wet early snow coats this hulk at the mouth of Chilkat Inlet. On the far shore is Davidson Glacier; Chilkat Inlet flows into Lynn Canal at this point.*

But for most of the gill-net fleet, the end of August is time to head north hundreds of miles to the forbidding glacial fjords of Lynn Canal and Stephens Passage.

All along the North Pacific rim millions of mature dog salmon, following some mysterious instinct, begin their final journey. Past Cape Ommaney and Cape Spencer they go, up Chatham and Icy straits, and into the big mainland rivers that wind down from Canada: the Chilkat and Chilkoot, the Taku and the Speel.

The fish run strong in the fall, and the price is usually high. It is big money; we double our season up there, or more, in a few short weeks. But the land and the water are cold and hostile.

After the equinox the weather changes—violently—for the worse. Stephens Passage fishing tapers off, and the hundreds of boats that were once scattered throughout the region jam into Chilkat Inlet at the head of Lynn Canal for the best fishing and the worst weather of the season. Each year, I look at my friends and our boats and hope that as many return as go.

*Trollers and packers ride rafted together in the fog at Juneau. When the weather breaks, the boats will head out to Lynn Canal.*

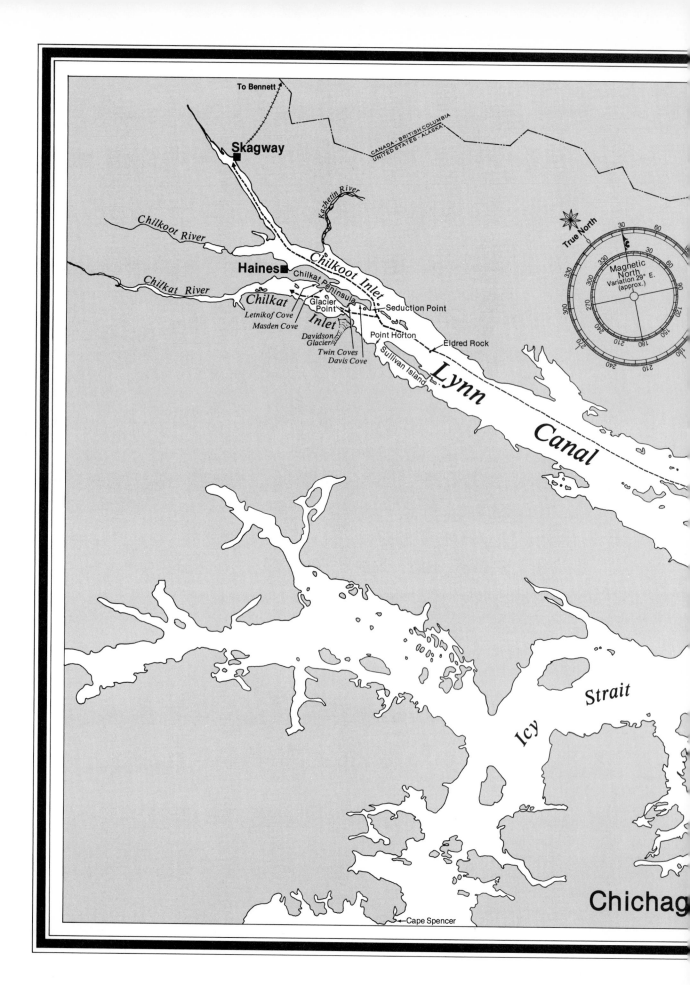

To Bennett

**Skagway**

CANADA · BRITISH COLUMBIA
UNITED STATES · ALASKA

Kazhelin River

Chilkoot River

**Haines**

Chilkoot Inlet

Chilkat River

*Chilkat*    Chilkat Peninsula

Glacier Point

Seduction Point

*Letnikof Cove*

*Masden Cove*

*Inlet*

Davidson Glacier

Point Horton

*Twin Coves*

*Davis Cove*

Sullivan Island

Eldred Rock

*Lynn*

*Canal*

True North

Magnetic
North
Variation 29° E.
(approx.)

Strait

*Icy*

Cape Spencer

Chichag

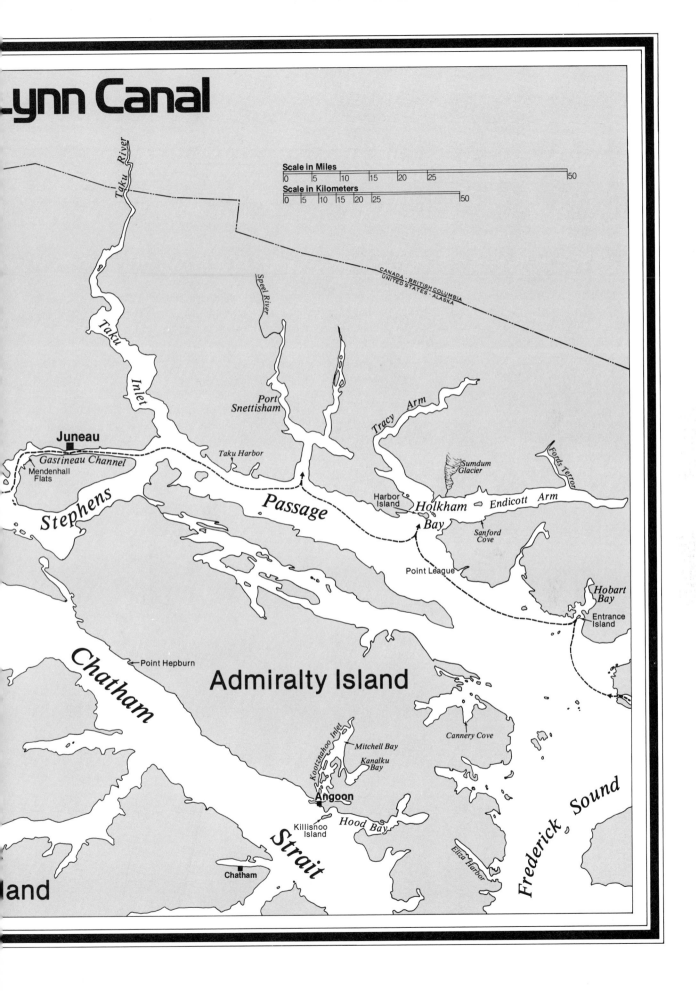

# Lynn Canal

Scale in Miles
0  5  10  15  20  25  50

Scale in Kilometers
0  5  10  15  20  25  50

Taku River

Speel River

CANADA · BRITISH COLUMBIA
UNITED STATES · ALASKA

Taku Inlet

**Juneau**

Port Snettisham

Tracy Arm

Fords Terror

Gastineau Channel

Mendenhall Flats

Taku Harbor

Sumdum Glacier

Harbor Island

*Holkham Bay*

Endicott Arm

**Stephens**

Passage

Sanford Cove

Point League

Point Hepburn

*Hobart Bay*

Entrance Island

**Admiralty Island**

**Chatham**

Cannery Cove

Kootznahoo Inlet

Mitchell Bay

Kanalku Bay

**Angoon**

Frederick Sound

Killisnoo Island

Hood Bay

**Strait**

Eliza Harbor

Chatham

land

*A*UGUST 29—Clear and northerly weather continues. I changed the oil and filters on the boat, and Susanna spaded up the garden and covered it with long streamers of kelp. She said Sam just lay watching, head on his paws, as if he knew we were leaving. Finally at 4 p.m. we closed up the cabin and putted out to the float with a box of warm clothes and another of books. Jerked the engines out of the skiffs, covered them with a tarp, and we were off. The *Kay II* and the *Osprey* joined us, and we all tied together with spring lines fore and aft. The afternoon was fair, the water still. The three engines throbbed and left a big wake boiling behind, and we all crowded into our pilothouse with a jug to while away the hours to town.

So we ended the summer season, running to town with a friend tied to either side, and the strait red with the dusk behind. Made the narrows at dusk, to pass through single file, the trees again close on all sides.

The big cruise ship, *Princess Patricia*, passed us in a tight spot, and we idled in the shallows to let her by. Just then spotlights were turned on to light up her two red stacks. In the brightly lit windows, we could see people in evening clothes looking curiously out at the black. A mile later, the Alaska ferry *Matanuska* passed, and when I looked astern, the two boats looked like two bright cities in some long dark valley. At 1 in the morning we made our turn into the tide, and tied to the city float in Petersburg.

*AUGUST 30*—Hit town with the lists this morning, to buy supplies for what could be a couple of weeks without seeing a town. Now all the gill-net districts to the south and west are closed for the season, and a steady stream of boats slips up the channel to fuel up and buy provisions. Each hour sees a few more, headed out and north into the vast reaches of Frederick Sound and the empty north country beyond.

*Below—Three boats lashed together and running to Petersburg. The stern view from atop Doreen's pilothouse gives a look into the trolling cockpit, with its own set of controls, behind the net reel. Companion boats* Kay II *and* Osprey *are left and right. Opposite—Seiner and packer together in Frederick Sound.*

"Between here and Juneau, almost a hundred miles to the north, there isn't another town or even a settlement."

All done by 3, we slipped our lines, and I wrote in the logbook, "Northbound with fair skies and high hopes." The houses and canneries of Petersburg slipped astern, and the wide northern districts opened up beyond the point. North of Petersburg, the country is very different from what we are used to in Sumner Strait. The passages are wider, the shores bolder, rising back to snow fields and glaciers; icebergs are a hazard the year round. Between here and Juneau, almost a hundred miles to the north, there isn't another town or even a settlement. For three years now we've headed north on such a fair August day. But how different our return weeks later! The weather's turned by then, and the fall gales have started their sweep up the coast. The trip that takes two days in August might take a week or more when we return south, and every mile we gain is fought for.

The night was starry, and we felt our way in to Entrance Island on Hobart Bay, where other northbound gill-netters crowded the sinking float. Here were friends we hadn't seen since the spring, and we talked, trying to figure out just where to go, for there are four or five areas open north of here. The fish hit each area at a different time, and much of our season depends on guessing and making the right move at the right time.

*AUGUST 31*—Followed the mountainous shore north on this last day of August, the water still, the sky clear. Port Snettisham opened up at 11 in the mountain wall of the mainland, and we passed inside. It's as good a place to start as any. A long deep arm of the sea, it winds some 15 miles into the interior. Its shores are for the most part steep, rising to peaks over a mile high. Almost two miles wide at the entrance, the canyon narrows to less than a quarter of a mile in the upper reaches, with sides so sheer that we passed into their shadow at 3 in the afternoon. We anchored in the basin at the head, not another boat in sight, and walked ashore past the ruins of an old water-powered sawmill, unwinding a little after the run and before tomorrow morning's fishing. There was frost on deck after dark tonight, and pale northern lights shimmering above. In the distance, the rush of the Speel River, tumbling down from the Canadian mountains to the east. After so long in easier country, this empty canyon awes us tonight.

*SEPTEMBER 1*—Pulled the anchor at 7 under a gray and mean sky, to cruise the inlet looking for fish: a jumper, or a splash or a swirl on the surface. But saw nothing at all, so moved to lower reaches, where a few boats had moved in from the strait beyond. Nothing there either, but we were there, so we just set the net off the nearest point when the period started at noon. The water is milky blue, but completely opaque from all the fine glacial silt, and the net disappeared completely behind as we set, leaving just the corks, stretching away from the beach in a long graceful curve.

In summer fishing, the water's clearer, and you can run your net, look down into the water to see if there are any fish in it. But here, not a chance, you couldn't even see through a glassful of this water. Watched that net for an hour, and saw just one wiggle. Many corks went down, but I thought it was just the new net tangling on the first set. Started to pick after a while, disappointed, and not expecting to get much, and had less than a fathom in when the first fish came over the rollers, blotched black and purple, dog salmon with long evil-looking teeth, alligators we call them. After that it was solid fish, coming over the roller six and seven at a time, tangled and twisted in the web. Susanna and I picked steadily, the net coming in between us, over 270 fish in that first set, our best this year, all dog salmon with a few scattered silvers, with a 10-pound average or better. Laid the net out again in the same spot, let it go for an hour, and picked for another 250. By then we were low in the water; another good haul and we might begin to have problems, but I was afraid to call on the radio for a tender after just eight hours' fishing, or the whole fleet would be in here. The rainy night settled in, and we laid out the net again, and hoped that a tender would chance our way.

*Left—Going north into Stephens Passage, we begin to see floating islands of ice. Below—At Bird Rocks, Cape Fanshaw is astern and the broad channel of Stephens Passage opens up ahead.*

On both sets, hardly a cork moved as the fish hit, and when I picked, each fish was already still and dead. Not a ripple on the surface, but under the boat a silent and determined mass of fish passed, headed for the river, and perhaps even for the Canadian lakes beyond, at the very end of their long journey.

As we drifted on our net tonight, it was dark, with only the faint light of the moon somewhere above the clouds, enough to barely show the jagged mountain ridge far above. My net light was just a dot in the vast blackness. I looked around and counted 11 boats in all of the inlet. At midnight, I spotlighted a passing tender to a stop and we pitched off almost 6,000 pounds of fish, our best day of the year.

*SEPTEMBER 2*—The fishing went down to 30 to 40 fish per haul after last midnight. Today was gray with rain, and outside in the strait, boats on the radio were complaining of the wind. A few more slipped into the inlet in the afternoon, but we still have it pretty much to ourselves. Night came very early, black with steady cold rain. But we have little worry about trash in the water here, or strong tides, so we can even sleep while the net's out. Bob called at midnight with a big score to the south, so we ran and set off his end for more than 200 fish in a short set.

With our hoods up, rain pouring down, Susanna and I needed almost two hours to pick our last set of the day, our world a circle of rain and water under the picking lights. I'd step on the pedal, another tangle of fish and net would come over the roller, we'd pick like madmen, throwing the fish forward, then roll on another fathom or two of net, and start all over again. Bob was out there somewhere, and others, but their lights were lost in the rain and mists. Every now and then, I'd duck into the cabin for a quick look at the radar screen, for outside there was no sky, no horizon, just inky black in all directions. I can hardly describe the blackness of these steep-sided northern inlets on such a night. After moonset, I turned out the masthead lights and stood out on deck for a few minutes in the rain. One's eye searches for a point of reference, a faint star, the dark horizon, but there was nothing—even my net light, a thousand feet away, was lost in the gloom—and it was an uneasy feeling.

*SEPTEMBER 3*—Susanna and I met the dawn this rainy cold morning, back in the stern picking fish after another big set. Our hands were numb from the

*Above—Wet, numbing cold and sore fingers. Susanna picks the net for dog salmon. Opposite—The welcome shelter of a cove on Harbor Island, Holkham Bay;* Doreen *rides at anchor in the bay.*

icy water. We wear gloves, but they're thin rubber, and soon shredded by the teeth of these dog salmon. But it was all worth it to slip alongside the tender at noon with another big load, for our biggest week this year. And from the looks of the packer, everyone's had a few fish, for she was deep in the water, decks almost awash, fish right up to the hatch cover. We were the last boat, and when we were done, the packer took off down the channel, pushing a big wall of water and headed for town, 12 hours away to the south.

It was a gamble coming in here this week, but it paid off. Now the older fishermen tell me that the hot spot next week will be up in Lynn Canal, 60 miles to the north, and I suppose we should go, but that's the windiest part of this whole country, and I sure hate to run from the kind of fishing we had this last week.

Ran two hours to the south in the afternoon, to Harbor Island in Holkham Bay. With four days to kill and no trolling in this district, it's our time to poke around and see some new country. Found a sheltered cove with empty buildings slowly falling down. In the woods we walked past row after row of cages, all empty. This place is miles from anywhere, yet the

house had been meticuously built, and in a corner we saw an old phonograph, still with records, gathering dust and mold. Somebody's dream is now lifeless and abandoned.

*SEPTEMBER 4*—The day came in shades of gray. The strait was empty and still. Rowed Sam ashore for a walk, but even he was subdued by the mood and stayed close to my side. At 9 we started up and slipped over the shallows to the east, past huge grounded icebergs, larger than three-story buildings, and entered Tracy Arm.

Even after all the country we've seen this spring, we were unprepared for the spectacle of this glacial canyon, winding for almost 25 miles into the interior through forbidding high country. In places the walls were almost vertical, bare wet rock without a piece of moss or a bush, and scarred for miles by the receding glaciers. Here and there side canyons stretched back to glaciers only recently receded from the salt water. At each bend in the channel, the country seemed wilder, the walls steeper, the fjord thicker with ice. Barely halfway to the head, we slowed to pick our way through, pushing the ice aside at times with our bow. It was no country for thin wooden boats. Here and there a few larger bergs towered above the rest, delicate blue with high ridges and dark ice caves, gulls and eagles sitting on the peaks. One berg must have been 800 feet

long and 90 feet above the water, and we passed it uneasily, wondering what sort of commotion there would be in this narrow inlet if something like that were to calve off one of the glaciers while we were there. Reached the head of the inlet at 1, and a wilder or more primeval place I have never seen. We lay in an ice-choked basin, perhaps a mile across. To the north and to the east, two massive glaciers, mottled green and white, disgorged into the arm. The face of each was perhaps half a mile wide and several hundred feet high, dominating the scene. The rest of the basin was bare rock walls, scarred and gouged. The air was filled with the rush of a dozen streams, and now and then a deep rumble, like nothing I had ever heard before, filled the air as millions of tons of ice creaked another inch or so toward the sea. The water itself seemed to vibrate; we could feel it through the hull, and looked up to see chips and huge blocks of ice break off the glacier and tumble into the water far below. The day was raw and cold, the whole scene evil and frightening. Susanna and Sam rowed about in the skiff, and I idled the boat close, following them as they moved through the ice.

Ran out at dusk, thinking that Tracy Arm was all I'd heard about it and more, too. I've been in a lot of lonely places, but never anywhere as plain hostile as where we were today. Ran back here to Harbor Island

again and snuggled in just as close to the trees as I could get the boat. The Indians say there are spirits up in the arm, and tonight I believe it.

*SEPTEMBER 5*—The day came cold and bright, a true fall day, and above the bay was the first fresh snow of the coming winter on the lower mountain slopes. I was all for pushing Sam over the side to swim ashore, but thought better of it, and after a while two deer appeared at the edge of the clearing to graze in the grass as we waited for the tide.

At 11 we ran for two hours down lonely Holkham Bay to Fords Terror, where, at 1, with little water under us, and at dead slow, we passed the rapids in the creek-like entrance. Hardly spoke a word for the next mile, so overpowering was the scenery. The channel was barely a hundred feet wide. To the north a sheer rock wall rose out of the water for a thousand feet before sloping back out of sight. To the south was a rocky beach rising rapidly to dark forests and snowy peaks above. Old John Muir was the first white man in here, around the turn of the century. He was so affected by the scenery he named it Yosemite Inlet, and I don't think there have been too many visitors since then. We passed a waterfall that was falling free at least a hundred feet into the trees below. The gorge opened up to a basin perhaps one mile by three, and we dropped the hook and walked, until our boat was just a white dot on the far shore. The lower slopes were forested, but above, the rock rose slick and bare, and I believe even a mountain goat would have trouble getting in here. We walked all afternoon, and never a sign of man.

The sun went over the mountain at 4:30, and the evening came early and chill. At dusk, flight after flight of ducks came in low and fast, to settle on the water near the shore with a rush of many wings and soft callings. During spring and fall in these lonely bays, we're witness to the mysterious drive of migrating birds. Once, on a May morning, I stood on the deck of a crab boat, a thousand miles to the northwest of here, speechless at the sight of millions of migrating birds pouring over the shoulder of the

*Doreen lies in the stillness at Fords Terror in Endicott Arm. The 7½-mile-long estuary is named for the 19th-Century sailor who found the creeklike entrance bar (overleaf) full of floating glacier ice on a falling tide.*

mountain and spreading out on the water like a dark carpet, all headed north to the Arctic.

The night was chilly, with northern lights again. Stood out on the deck and watched the birds with Susanna until the cold drove us inside. Yesterday and today, the places we visited made us feel tiny indeed.

*SEPTEMBER 6*—First frost! The stove went out in the night, and we woke to find Sam nestled in between us. To go out on the frosty deck on such a morning, with the still glassy basin around, and dark forests and frozen hills above—words can't tell it, pictures can't show it.

Idled out slowly, taking it all in again, made the entrance at slack water, and ran a dozen easy miles to Sanford Cove. This is our third day now without seeing another boat or another person. Here at Sanford Cove rotting pilings stand on the beach, and skeletons of buildings crumble in the woods, remnants of a cannery or a mine, who knows? There are two streams here, with good water and plenty of gravel for spawning, but not a sign of any fish, a discouraging sight. Walked in the woods until dark, Sam growling twice, hackles up; I suspect that a bear was close. Tonight, while rowing out to the boat, I rested on my oars and just then two honkers, Canada geese, passed over our heads, talking to each other before they decided on a place and landed. Had clams from the beach for dinner tonight. Now we're just sitting here, fresh-picked flowers on the table, Sam asleep at our feet, not a soul or a light for miles, and our cup seems pretty full.

*SEPTEMBER 7*—Deep in the night I woke to a noise and found a bathtub-size chunk of ice scraping past the boat in the tide, illuminated by the northern lights blazing above. In the morning, there was Sumdum Glacier across the way on the mainland shore, shining and white above the trees. We ran through rippled waters in the afternoon to Port Snettisham where we found the *Cape Hason* and others. They had just stayed there, whiling away the days, canning fish, panning a little gold and walking the ridges for mountain goats. Talked to them about Fords Terror, and no one had been there, or even knew of anyone who had. Gill-netting again tomorrow, and boats wandered all evening into the anchorage.

*SEPTEMBER 8*—Up at 6 on another cloudless morning to run out to the point where I wanted to make my first set. The fishing period didn't open until

noon, but I knew after last week that many more boats would be in here, and if I wanted a particular set, I'd better get there early and stay on the set until fishing started. Sure enough, boats poured into the inlet all morning until noon, and I counted 40 boats where 20 would be a crowd. We had just 10 fish on the first set, so all that waiting was for nothing. Moved to the south shore for the same results, so we picked up and ran, out of the inlet and south in Stephens Passage; I couldn't see staying for that kind of fishing. There's a waterfall in Stephens Passage, and I set off it and let the tide sweep us north along the beach and around the corner into the mouth of the inlet. Picked up for 40, nothing spectacular, but good fishing, so ran back and did it again. Bob came out after a while, then his uncle, but beyond that we were alone. So the day passed. Nothing big, but just steady fishing, and that's what pays the bills. With enough tide to keep the three of us apart and bring in a few fish, that section of beach was just right. On the big radio, boats complained of the poor

fishing, and after a while we just turned it off and talked among ourselves on the low-powered CB's. The shore there is bold and deep right up to the beach. When night came, the falls were our guide—we'd run in and set the net right at the falls. Then we could even take a nap as the tide slowly moved us down the beach, and when the roar of the falls grew faint, it was time to pick up and run back. Never fished in a more enjoyable spot or had easier fishing.

Silver salmon often run out in the strait at night, so we moved out at midnight to try a set. The water was glassy, the whole of Stephens Passage was silent and empty. Nowhere could we see another light.

*SEPTEMBER 9*—Same thing, same spot. I know every rock on that beach now, and I back the boat right in to the beach before throwing the end of the net over. Once I thought Sam was going to jump ashore and not even get his feet wet. Tonight at dusk the tender came by to take our fish, and a crewman was surprised at our loads. He said the fishing was

*Left—Out in Stephens Passage, gill-netters drift with the tide, their nets floating free on a dead calm day. Below—Off Stephens Passage, the packer* Annette *waits for her boats in Port Snettisham. On a rough day, the protection that these landlocked waterways offer to packers and fish boats unloading is a godsend.*

pretty poor in all of the district but that they had hit it big in Lynn Canal. As we lay there, the mists settled in on the hills around us, then the tender was gone, off to pick up fish from the boats in the inlet. We've never had fall fishing like this before—last week and now this—good fishing and good weather to boot, our three boats here with this spot all to ourselves. Enjoy it while we can, I guess, for next week it's Lynn Canal for sure and that's always a zoo.

*SEPTEMBER 10*—Almost got caught with our pants down last night. The norther started up with a rush at 2 a.m., the first gust easily 25 knots, and us off a lee shore with all the net still out. I put that reel drive in high gear and never stopped hauling until the whole net was on, rolled it all in, fish and all, with the water breaking white just off the stern. Another two minutes and I would have had to cut the net off and run. Moved offshore and set again, but the short chop was uncomfortable, and the wind was coming up all the time, so I hauled back and ran. The stars were lost in the clouds and there was just the faintest gleam from the snow above. We ran up to the very head of the inlet, the dark walls closing around us and only the radar screen showing the channel ahead. Fished there through the dawn for just a handful, and let my third set go until noon, the tide and river sweeping us around the points and down toward the tender. Above, there was fresh heavy snow on the hills, closer to the water now. Noon came and another week was over; we've been lucky and I know it. Another week or so and we'll have the winter in our back pocket; we've scratched out another season even though it all looked so gloomy for a while.

Sold our fish and started for Juneau, but outside the inlet, the northerly was a full gale; had our fill in just an hour, so ducked into Taku Harbor where we lay with doubled-up lines and the wind howling in the trees on shore and in the rigging. The *Cape Hason* was there, and others; we all crowded in to drink and tell lies about how bad we did.

*SEPTEMBER 11*—Taku Harbor. Northerly gale continued all day. Big combination gillnetter-troller tried to get out this morning, but it was back in an hour, decks washed clean. We walked the beach in the afternoon past the usual cannery ruins, the creek heavy with spawning dog salmon. Sam waited for them in the shallows, then chased them up the narrow

reaches, and even dragged two out and left them flapping on the beach. There was an old gill-netter on the beach, full of holes and rotten, but I salvaged four brass portholes, dark and moss-covered, and with a little polish they came clean. At 4 p.m. the tide turned against the wind outside, and the water was white for as far as we could see.

*SEPTEMBER 12*—Woke twice in the night as vicious gusts buffeted us at the float, working the boats back and forth against the lines. But the day came bright and clear, with the wind down to a steady 25 knots. It was a chance, and we tried it, for a very tricky ride across Taku Inlet. The cold northeast wind, sweeping down from the glaciers and cold inland country to the east, raised a fearful short chop which I found unusually difficult to traverse. Couldn't quarter across at all. Jockeyed the wheel and the throttle for

*Bucking a norther in Taku Inlet, putting spray over the bow.*

three hours up to the lee of the land on the far side, then ran down to Gastineau Channel. Even there, with the high buildings of Juneau in the distance, the wind swept down from the ridges in gusts, and heeled us over. Reached the harbor, finally, where we tied up seventh boat out at a very crowded float. Left the dog on the boat and treated ourselves to a hotel, a hot bath and a night on the town, where we continually bumped into gill-netters in all the fanciest places.

*SEPTEMBER 13*—The northerly gale continues. Back to the boat this morning to find that the angry dog had shredded trash all over everything. Retaliated by giving him a needed shampoo. The harbor's almost solid with boats now, waiting for a break in the weather to run north to Lynn Canal. Went out to the end of the road in the afternoon to have a look at it. Where we stood, in the lee of the land, it was warm. But out beyond the point was a wild sight indeed. The wind was a steady 50 knots, with higher gusts. A big tide was pouring in from the south, and the canal was a confused mass of white water, overfalls and breakers from shore to shore. We just looked and didn't say too much. Yesterday a boat and crew were lost out there; farther up the canal, three boats—swamp-loaded with fish—were overcome by the seas and sank, the crews lucky to get off with their lives.

*SEPTEMBER 14*—The day came foggy, but it burned off in an hour, and the sky was bright. The wind had gone in the night. By 8 there was a parade, past the fuel dock, and on to Mendenhall Flats, where we crept through the weeds and tall grass to save 40 miles. The tide was high, but even so we bumped and chewed our way through in places. At 9 we passed Point Lena, where the last road and house fell behind, and the wild lonely canyon on Lynn Canal opened up ahead.

The water was so flat we could have made it in a canoe today. But of all the places we gill-net, this is the one we like least. Chatham Strait and Lynn Canal form a kind of wind tunnel, almost 200 miles long. In the fall and winter, it sucks the southeaster in from the ocean and the northerly down off the frozen interior. Sometimes the wind will blow for weeks without

*"Hold still, Sam, we'll be done in a minute." Sam gets his shampoo.*

*Above—Fishermen's picnic ashore at Davis Cove, Chilkat Inlet. Right—Chilkat Inlet displays the somberness and the power of these northern districts in the fall.*

stopping. One fall we waited at Skagway for a week in a southeast gale, the seas topping the breakwater on the big tides as we lay behind it with doubled-up lines. Another time a friend took off southbound in calm weather, but when he was halfway down the canal, that norther started up, and in just a couple of hours he was burying his bow at the bottom of each wave. He nailed his hatch shut, put on his life vest, and made harbor with the water halfway up his sputtering engine, after the dirtiest kind of afternoon.

Dropped the hook at Davis Cove in Chilkat Inlet for a picnic ashore with three other couples. The dogs played, and we built a fire on the beach by an old trapper's cabin, as boat after boat passed into the inlet beyond. Across the water was a steep mountain wall, split in places where dirty white glaciers curved down from the year-round snow fields above. Twice tonight, as we sat eating, the glaciers boomed and cracked,

filling the air with their thunder, and we looked up at them six miles away and saw ice and snow tumbling down their faces into the trees below.

*SEPTEMBER 15*—The day came foggy with a lot of engines close as many boats jockeyed for a set four hours before the opening. Just north of here are Chilkat Inlet and the Chilkat River, with its tens of thousands of acres of spawning grounds. By the time the fish are this far up the canal, they're milling around by the thousands, waiting for the right tide or whatever it is that makes them start their last drive up the river to spawn. The water is deep and our nets shallow, but now and again the fish rise in great masses, and you can sink your net with fish in just a few minutes, with boats fishing on all sides of you. Two years ago I was set in the inlet and the man next to me, 50 yards away, got a shot of fish like that—he must have had over 500 in that one set, and he picked for most of the afternoon. I was wild, I tell you; I set on all sides of him, but they just came up in that one spot,

and he was there. So we all crowd into this narrow arm off the canal, 200 or 300 boats or more, and this remote inlet, empty and still for most of the year, becomes a city of boats for the few weeks when the fish run.

We started on the beach, just as the fog lifted, for 110 fish, but the boat outside of me doubled that. Sometimes we can fish where we want, but most likely we fish where there is room, at times running for an hour simply to find a spot to put the net in. Fog settled in again with the night. With boats and nets on all sides, we just fished the one spot, between a pinnacle rock offshore and the beach, never a big set but steady good fishing all night, 60 and 70 fish a haul.

*SEPTEMBER 16*—Fished until noon, the end of the period, in the thickest kind of fog. There were engines and occasionally voices close, but I never saw a thing; the end of my net disappeared into the white, 30 feet off my stern. Then, about noon, it all burned off at once and I looked around to see hundreds of boats headed across the glassy waters to Seduction Point,

*Clockwise from upper left—Chuck Zieske hauls his net by hand, as hundreds of boats crowd into narrow Chilkat Inlet. Packers load fish in the shadow of the glaciers at Seduction Point. Crews have a party while waiting for their boats' turn to unload.*

where a dozen big tenders lay at anchor. The afternoon was fair and warm, with silent mountains and glaciers above, and we lay deep in the water with our load, talking with friends as they passed, drinking beer and waiting our turn to unload. When it came, the tender was already deck-loaded with fish, right up to the tops of the bulwarks, all the way up to the bow and around the anchor winch. There'd been still weather and good fishing for everyone; this was Lynn Canal at its best.

Headed up the canal in the late afternoon, a big check in my pocket, and didn't it feel good! We were running for Skagway with a couple of days to kill. To starboard were the Kazhetin River flats; already the birch and alder are turning yellow and gold. We

passed the town of Haines, and the canal narrowed, the walls growing steeper for the last miles. We tied up behind the breakwater in Skagway at 5 in the afternoon. Skagway is the end of the line. Just to the north, the canal ends in a shallow inlet, the mountains like walls on all sides. Skagway is different from any other town in the region—a row of old false-front stores and houses sprawled across the valley between the mountains. Even the climate is different, with hotter summers and colder winters. Skagway's heyday was the 1898 gold rush, when ships dropped loads of prospectors to start their long journey over the mountains to the Yukon River and the Klondike gold fields beyond. Now Skagway gets by on the business of the tourists from the cruise ships and ferries in the summertime, and it picks up a little railroad business year-round from a narrow-gauge railway that runs over the mountains to Canada.

*SEPTEMBER 17*—The day came bright with frost on the float. Locked the dog in the boat and headed for the train for a sightseeing trip. Very congenial group of fishermen in the last car, with much storytelling and sipping from brown paper bags. The few tourists quickly moved to the next car. The train started with a jerk, and soon we were winding through vertical country, snow fields above, a deep gorge below, the town and the canal lost behind. We sat and watched it all slip past, but the prospectors had to do it all on foot or horseback, over the snowy passes with their heavy loads. Stopped for lunch at Bennett in a cavernous dining hall miles from the nearest settlement. This was in Canada; a few miles farther away was the border of the Yukon Territory, but Bennett was as far as the train went on this excursion.

Had a long look at the photos on the walls after lunch; they showed a tent city of prospectors on the edge of the ice. Lean and hungry-looking men cut down the trees, hand-sawed them into planks to build boats, and waited for the ice to go from the river so they could start their long journey. What the lure of gold will do! After looking at those photos, I wondered how many of those men got what they were after.

On the way back, the train was stopped by a snowslide across the tracks, and we got out to stretch

our legs and wait for the plow. The conductor said pretty soon they'll have to use the big rotary plows to get the trains through at all. Finally got under way, the countryside rolling past, Susanna's head on my shoulder. Who met us at the station barking? You guessed it. Got out the window, I guess.

*SEPTEMBER 18-20*—Indian summer in Skagway. A little work on the net, a picnic, a hike up to the lake, a trip uptown in the evenings: we've mostly been doing whatever the days suggested. This little town at the very head of the canal has always been special for us. We get here at the end of a long season that started, perhaps as far back as April, in the waters of Washington State, or at Noyes Island in the Gulf of Alaska. We've already made our money, if we're going to, by the time we get here, so Skagway's just a vacation. We wait there for the next fishing period, maybe hoping that it will be the last, so we can point the bow south and head back to wherever we came from.

*SEPTEMBER 21*—Under way at the first pale light, down the canal as the sun came fair again over the mountains. Anchored at Glacier Point and hiked through waist-high grass and stands of alder and birch. There's an old trail, and we followed it to where Davidson Glacier, after coming so proudly through the mountains, ends in a sad little pool surrounded by mudflats. Saw many signs of bears—fresh tracks and droppings everywhere. I was unarmed and uneasy, but Susanna pooh-poohed my nervousness. Sam was undeniably happy, sniffing everything, running across the flats, splashing across the water. On the way back he was just a black tail winding through the field of grass ahead of us. The sky was pale blue, the meadow stretched away to the horizon, and from where we walked, the canal was just a silver ribbon winding away to the south. There's a ring around the moon tonight, so I suspect a change for the fishing period tomorrow.

*SEPTEMBER 22*—My suspicion was right. The wind was up to a half-gale from the northwest, but boats were slipping out of the anchorage at 7 to jog on a set and wait for noon. Started fishing in the inlet for

50 fish on the first set, but the area was too crowded, even for me. Counted 138 boats inside the inlet, in an area two miles by four miles, each boat with a net a quarter of a mile long. Moved outside for good fishing, but it was very wet work in increasing winds. The night came quickly with racing clouds, and blacker than black, the wind a steady 45 knots with higher gusts. Almost got blown onto the beach while I was picking up the net; all of a sudden there it was, the gleam of white water, with the trees looming up beyond. I rolled the last 15 fathoms of net onto the boat without stopping, and backed out, almost burying the stern and us, too, in the short chop, but there was no room or time to turn the boat around. Set again farther out, but the weather was too much, the seas coming over the stern as we picked. Susanna and I were soaked inside our oilskins.

Gave it up after midnight and jogged through combing seas in the black to the tender in the lee of Point Horton. Even in there the night was wild. The tender skipper said Bob's uncle's boat had sunk; it filled and was gone with hardly even time for a call on

*Fishing in Lynn Canal at night: Right.—Susanna picks the fish from the net. Below—A packer and a gill-netter unload fish.*

the radio. The old man and his son were two hours in a little skiff, paddling and bailing that icy water with their hands before somebody picked them up, about half dead from exposure. Several other boats were blown ashore, but everyone was safe. Our anchor took on the fifth try; there are boats on all sides of us, and the anchorage is poor. The bottom is just a steeply sloping shelf, dropping off quickly to 80 fathoms. If you anchor too close to the shore, the tide will set you on the beach when it turns; if you anchor too far out, you'll drag off into the deep. With the spotlight I picked out two boats up on the beach and laid over on their sides with the falling tide. Chuck came alongside, glad to find someone with good anchor gear. He said Doug's *Anatevka* had just gone on the beach. The winds are violent, even in here, but our anchor seems to be holding. Few boats left fishing.

*SEPTEMBER 23*—After an uneasy sleep, with the wind a constant whine, I was up at 4 to replace a spring line parted in a violent gust. Off at first light to fish inside the island, but even in those sheltered waters, the wind was a steady 35, and we had to jog off the end of the net to avoid pulling it out of shape from the drifting of the boat. Twice during the day a big red Coast Guard helicopter flew over looking for people on the beach. How it ever landed in that wind, I don't know. Finally, at dusk, the wind eased off to a steady 20.

*SEPTEMBER 24*—Small codfish moved in last night—I must have had to pick 500 of the soft-fleshed little critters on my dark set, mixed in with 120 dog salmon. Morning came with rising winds and southeast gale warnings, so I quit an hour early, sold and ran for Haines before it got any worse. The harbor was already crowded when we got there; we were lucky to get a berth on the inside. There was a fresh gale, as predicted, in the late afternoon, driving in a chop even behind the breakwater. The boats on the outside of the floats pitched and heaved, chewing through lines and fenders. Saw Doug in the bar this evening looking a little shaken. He said he'd known he was close to the beach, but figured he was O.K. as long as he could see the fishing lights of boats to leeward of him. When he hit the beach, the first sea swung him broadside, the next drove him farther up and started to fill the cabin. By then there was nothing to do but grab his sleeping bag and try and make it through the surf to shore. It

was then he realized that those lights to leeward were two other boats, also blown ashore.

*SEPTEMBER 25-29*—Lying inside the breakwater at Haines. Southerly gale continues day and night with driving rain. Even on the inside, the surge is uncomfortable at times. We've hit every bar, seen every movie. The town is very small, and the waiting is hard. The money is good up here, even when we're fishing just the one day a week, but the waiting sure is tedious at times. I don't like lying in the bunk night after night—with the wind howling in the rigging and driving the seas against the breakwater—worrying about that long thousand miles ahead of us. Each year we get a good week or two up here, but then the weather turns, and we all get itchy for the weather to break and the season to end.

*Below—Houses on the old Army post at Port Chilkoot near Haines. Overleaf—A clutch of boats waiting for good weather in Chilkat Inlet as a gale sweeps past.*

*SEPTEMBER 30*—The clouds lifted this morning to show fresh, heavy snow on the hills all around, closer to the water than I've ever seen it at this time of year. I remember an old fisherman's advice about fishing Lynn Canal. He said to run south as soon as you saw the first snow on the hills. Even if you're right in the middle of a big set, pick up that net and start running. But greed beats caution, I guess, because we're still here.

Moved to Masden Cove in Chilkat Inlet in the afternoon, and had our first dry walk in a week, Sam's ears flapping in the wind and Susanna and me leaning into it. The wind came on hard from the north just at dusk. We'll be fishing tomorrow, but I don't look forward to it if this weather keeps up.

*OCTOBER 1*—Had a sleepless night; we dragged anchor twice. Finally had it set and was just settling back into the bunk when wham! a big gill-netter dragged down on top of us, and I was wide awake again. The dawn came cold and windy, with Sam's water bowl frozen solid. Started out at noon in the inlet with the usual big crowd on all sides, but it was very sloppy going. Good fishing, but the reel drive was straining to pick up the net against the wind, and the corkline was coming in tight as a bar. Started taking solid water over the stern, so moved inside the partial lee of Sullivan Island after dark. At 10 the stars disappeared behind heavy clouds, and an hour later a freezing rain started. That southern trail looks better and better all the time.

*OCTOBER 2*—The wind continued northwest, gale force, sucking down off the vast cold mainland to the north. After midnight last night the sleet turned to snow, blotting out the lights of the boats all around. Gave up jogging on the gear at 3 a.m.; had to make three tries to pick up the end of the net after my last set, and was soaked inside my oilskins again. I let that set go for four hours, hoping the weather would get better. But it didn't, so I picked it in driving snow, with ice building up on the corkline. But we had almost a hundred fish, so I ran back and set the net out again. Susanna stayed inside on these, and I didn't blame her

*Below—The gill-netter* Meredith E *bucks into a southeast gale in Lynn Canal. Right—Tenders take refuge from the storm at Twin Coves.*

*Snowbound in the Twin Coves anchorage in Lynn Canal, Allan reaches into his refrigerator for a can of pop.*

much, either. Picked the last set at noon in very strong winds, the reel drive barely able to pull the net in. Had zero visibility in a snow squall that was stinging my face, and I had to turn away and let the squall pass. Even on the inside, where we were, the wind was blowing the tops off the seas, and they were slopping heavily into the cockpit. I was glad to get the net in, put on dry clothes, and jog up to the tender. It was heavy going, running at reduced speed with little visibility. The wipers froze solid, and we started making ice on the rigging.

Bob called from up in the inlet, said not even to try and get up there—he was jogging in a snow squall, his radar iced over and useless; he was hoping the weather would let up enough for him to find the harbor. Detected some anxiety in his voice, and didn't blame him much either, with his wife, two little girls and a sick uncle on board.

It took us three hours' bucking before the land came out of the white and we saw the coves where the tenders lay snuggled up into the trees, their deck lights on in midafternoon, the squalls blotting out the hills beyond. We were at Twin Coves on the end of the Chilkat Peninsula. We were the last boat at our tender; we unloaded and the tender was gone, lost almost at once in the swirling snow and gloom. Our week's over but the tender still has almost 200 miles back to the cannery. We ran up to the head of the cove and tied alongside three friends with all our anchors out ahead. Night came quickly, with 15 boats crowded into this tiny anchorage. The weather's wild out there, but cozy enough in here, with dinner on the fry,

Susanna here beside me, Sam snoring on the floor. Violent squalls again at 9, laying us all over with their power, our lines and fenders creaking. On the radio we hear that even the largest tenders are running for shelter.

*OCTOBER 3*—Up at 2:30 to watch the anchor in another violent squall. Boats ahead of us are reanchoring, their spotlights making white shafts in the swirling snow. Finally went back to an uneasy sleep with the wind shaking the rigging and snow hissing against the hull.

Dawn came gray and cold, the shores white under low thick skies. Heard tales of dragged anchors and parted lines on the radio. The northerly continued all day, at times fearsome in its power, snow blotting out the cove and the wind whipping the water white even in here. A good day for cards and visiting. Shoveled off the deck three times today. Never thought I'd be using my ice shovel to shovel snow. Think we'll just give it up and run if the weather ever lifts; this northerly's just about killed my desire to stay up here any longer.

Listened to station reports in the afternoon. The lighthouse at Eldred Rock, 10 miles to the south, reports 65-knot winds and 14-foot seas. The forecast is for southerly gales in the morning. Here at Twin Coves we're exposed to the south; we'll have to pick up and run as soon as the weather breaks, even if it's the middle of the night. Bob called from up in the cove in the afternoon. Said the northerly is breaking up the floats all around him. Told us to watch out for the southerly—it can switch around in just a few minutes.

*Above—Snowman by the small-boat harbor at Haines. Left—*Arctic Tern *runs through the snow in Lynn Canal, following* Doreen *to port at Haines.*

*OCTOBER 4*—Day came snowy but still. I led a single file of boats to town, with hardly 50 feet of visibility at times. We went south around Seduction Point and then north to Haines. We passed the bow of a sunken gill-netter, awash off an icy beach. Our wipers were frozen, the pilothouse windows were opaque with snow, and we had only the radar to lead us in. In the harbor, the floats were under two feet of snow, and boats were coming in from all the anchorages. Left the dog with friends, tied up the boat with five lines, and took a room in a hotel, with steam heat and a bath down the hall, feeling as if we'd wait until spring.

*OCTOBER 5-7*—Back on the boat, and there's a southeast gale driving around the breakwater again. Ran out new bow and stern lines; the old ones had chafed through in just a week. Old-timers say it can go on like this for weeks at a time. In the afternoon, I

hitchhiked out across the peninsula to Letnikof Cove, where Bob and a few others lay with tripled-up lines. It was a grim and desolate sight. The float was broken in half, part of it washed ashore alongside a boat. Another boat lay at the end of the float, only its rigging above the water. Beyond all, the cold, wind-whipped waters of the inlet.

*OCTOBER 8*—Northerly wind again, but clear. Borrowed a truck in the afternoon and drove out the road along the river, where all the streams were full of spawning dog salmon, and the trees above were full of eagles waiting for the salmon to die. Sam charged through the shallow water, driving the fish before him like a wall. It's six days now that we've been waiting for good weather.

*OCTOBER 9*—Alarm went off at 7. Still dark outside, but the wind was down. Fishing was open again today for just 24 hours. We should have bagged the whole thing and run while there was a chance to get out of this windy hole. Still, the fishing's been good up to now and $1,000 looks pretty good about the middle of February. So we stayed to fish, and after an hour the wind came up and it was the same old thing

*Above—A boat sunk at the float at Letnikof Cove with just its rigging above water. Right—Bob Anderson pulls the anchor.*

again, but very poor fishing—only 11 dog salmon on my first set, the worst I've ever had in the canal. Moved inside the island for the next one—7 this time. Had hardly any desire left to put the net in again, but we'd missed our chance to run, so decided I might as well put in the time and hope for a lucky set.

Picked codfish and jellyfish for two hours for 9 measly salmon. Rolled the net on for the last time and headed for the anchorage.

So, on a black night with gale-force winds and icy driving rain, our season has finally ended. We've a thousand miles ahead of us, and the weather is bad, but we've paid all our bills and have enough for the winter ahead and more. We talked about and looked forward to this moment for weeks, but when it came, it passed almost without notice. Susanna was asleep, and I dropped and set the anchor, and sat up with a cup of tea, listening to the wind in the rigging for a bit before I turned in.

*"We left at dawn to sell and be on our way south. . . . In a few hours the packers would be gone, too, and this bay would hardly see another boat until half a year had passed."*

# Home

It's a thousand lonely miles from upper Lynn Canal to Seattle. When we catch the last fish and our season is done, the weather has turned mean; the journey that was so mild and easy in the spring is a battle against poor weather and short days in the fall.

Through August and September the packers and seiners and gill-netters from the areas north and west of us—Kodiak, Prince William Sound, Cook Inlet and Bristol Bay—trickle through Southeastern Alaska on their way south down the Inside Passage. We hear them on the radio, headed down the line, their season over, their fishing areas closed.

Some quirk of geography and nature makes the Lynn Canal run the last salmon run on the coast north of Vancouver Island. The big Canadian runs are over by then, the canneries closed for the winter. When the fishing ends in Lynn Canal, perhaps a hundred boats head out at once. But the weather and the country swallow them up and scatter

*Eldred Rock in Lynn Canal,
the first landmark ticked
off the list on the long
trip home.*

them, and after a day or two, the boats are running singly or in small groups of twos and threes.

In a windy year, the trip home is a whole other saga. The weather is big, too big for little boats to fight, and nothing can be done but lying in and waiting when the wind blows. The trip is long and wearying, even hazardous at times, but there's a satisfaction to it. It's no small thing to finish the season in the big Lynn and fight south, up against it day after day, seeing hardly another person or boat for 10 days or more, finally coming through the narrow lower channels to homes and towns again. You can tie up the boat and go ashore for a long while, the season done, the winter made.

*It's either bucking weather
or waiting (overleaf),
or so it seems.*

OCTOBER 10—The day came raw and cold. We left at dawn to sell and be on our way south. I looked back as we finally headed out around the point; three packers lay in the bay, their deck lights illuminating the whole anchorage. Above, the bleak mountains rose up from the water, and the first cold light of day outlined the clouds. In a few hours the packers would be gone, too, and this bay would hardly see another boat until half a year had passed. As we bucked south we passed Eldred Rock, rising bleak and stark from the water, and I was glad to put it behind.

There were boats ahead and astern, all heading south, but soon they were lost in the cold rain of the gray day. Bucked south all afternoon. At dusk the wind came up, 30 knots out of the south, for a last dirty hour before we made Taku Harbor. On the shore, not a light; in the harbor, not a boat. The others had stopped in Juneau for fuel, and then decided to stay when the wind came up.

OCTOBER 11—The day started at 2 a.m. as the bow line parted in a violent gust of wind. We swung broadside for a minute; then the strain was too much, and the stern line went, too, and off we went. Three-quarter-inch lines, and barely a week old. Ran up to the farthest corner of that bay to anchor right under the lee of the shore. Even there the wind was strong. I sat up to watch as the anchor line came tight in the squalls, creaking against the roller. On the radio, a crewman on a packer 30 miles to the south said the packer was giving it up and running for shelter. I know the boat. She's 80 feet long and built for the meanest weather.

The day came with light airs and only a left-over swell, so we were off at 8, the strait empty except for a big packer on the far shore. Another front came through just as we were at Point League, and we spent an hour I'd not like to repeat. The first gust was a solid 40 knots, blowing off the tops of the waves, and it came right at the tide change. That's a steep and bold shore there, and we jogged just 100 yards off it. Outside of us the tide ran against the wind, and the rip boiled in overfalls and combers. On the beach, the seas threw solid spray up into the trees. Not much choice but to idle through with the tide behind us and our fingers crossed. It took an hour to make barely a mile. Susanna was right there beside me, saying little. Made Entrance Island at dusk, bucking against the tide

right up to the mouth of the harbor. It was one of those times when you just get caught and there's little to do but keep going. On the beach outside the harbor lay a fine 40-foot gill-netter, her bottom stove in, driven ashore in last week's blow.

OCTOBER 12—Down Frederick Sound under low skies, and not another boat in sight, the shores lost in the overcast. Then Rocky Pass with raft after raft of ducks trading back and forth in the shallows. It felt good to be running back near the trees again. The dusk came early and in the black we crossed Sumner Strait, the old stomping ground, with a swell driving up from the unseen ocean to the southwest. At 10 in the rainy, still evening, we passed into the cove and tied to our own float after some six weeks away. The skiffs were both awash, but I pumped one out and we rowed ashore to the cabin. I built a fire and Susanna and I sat there for a long while, looking at the fire and out at the dark night beyond.

We ran three days to get back here, and spent each night in some lonely cove with ruins on the beach. All that time, we hardly saw another boat. To slip in here on a black night with no other light in sight anywhere, after a journey like that, made the land seem more vast and empty than ever before.

*Canadian drum seiners (the boat on the right is aluminum) at the Indian town of Alert Bay.*

everything down for the ride, and off we went. Three hours with the tide, and then it turned. For three dirty hours there was water flying, shaking the boat on every sea. Reached Sunderland Channel, finally, and we were glad to be in a narrow channel and leave those wide waters behind for a day or two. In the afternoon we had constant squalls, rain and sleet and the night came black and early. We poked our way into Forward Harbour, just this side of Whirlpool Rapids, and tied to a log boom, with doubled-up lines around the slippery logs. Then, in that lonely spot, we had a turkey in Dick's boat that would have done credit to a New England grandmother. Oyster and mushroom stuffing, sweet potatoes, hot sticky rolls, salad, and pumpkin pie: we all brought something. But all hats were off to Dick, for he cooked and stuffed the turkey today out there in that foul weather. His was the only oven big

enough. He had to stuff the turkey two times; the first time, the boat took a roll and the whole works went onto the floor. How we ate! When we stumbled back across the slippery logs to our own boat, I was so full it looked like I was pregnant.

*NOVEMBER 8*—Off at dawn again to boil through Whirlpool Rapids with a 5-knot tide behind us, through overfalls and white water in that narrow spot. This is my favorite part of the trip, winding through these back channels—Wellbore Channel, Chancellor Channel, Mayne Passage, Cordero Channel—and usually there's less than a mile of water between the mountainous shores. Here and there a tiny settlement or deserted house—damn, I'd like to stop and spend a few days here, just poking around, but the weather's turned favorable and we'd better keep going while we can. I keep telling myself that some year I'll just bag it, and spend the whole season up here in British Columbia in a little boat, maybe a sloop or something, poking around. But then each spring comes and I have to go fishing.

Yuculta Rapids at slack water. We can miss the tide in those other places, but not there. Once I was late by an hour, but went through anyway in a big tide. That was enough for me; the current came close to putting me on the rocks despite all I could do.

We passed the entrance to Toba Inlet, which winds through the snowy hills to the northwest, and then the weirdly shaped hills and islands of Desolation Sound opened up to the south. Bob went on, but Dick and I stopped at the old cannery at Redonda Bay, just to walk again past the old buildings in the bright fall afternoon. We get our fall in bits and pieces, a few days of Indian summer in Lynn Canal, an afternoon here, maybe a week of good weather down in Puget Sound. Then, by the time we tie up the boat and fly north to the cabin, the long dreary winter has started.

We passed out of Desolation Sound at dusk, behind us all the land and channels dark. Ahead, the lights of the mill and the town of Powell River, and as we passed we could hear the beat of machinery from three miles away. So, after weeks of running through lonely country with only the smallest of settlements, we came out of the narrow channels to pass the first big mainland town. The noise and the lights were overwhelming. We slipped into the harbor at Westview in the black, past row after row of silent plastic runabouts, to lie beside a Canadian seiner, already freshly painted, ready for spring. Bob was there with an old Canadian fisherman, who eyed our tired-looking boats, rust-streaked and scarred, booms and skiffs double lashed, and said, "Had a long season, eh?" Tonight, for the first time in months, we went to sleep with the rush of traffic in the streets above, the night shift going to the mill.

*NOVEMBER 9*—Under way at 4:30. Malaspina Strait and dark empty shores on all sides again as a gem of another bright fall day started in the east. It was a fine morning, and the water rippled with the light northerly breeze. Our paths divided in the Strait of Georgia, and Bob and Dick took off for home across the U.S. border, just 30 miles away.

We could have pushed south, too, maybe made it halfway to Seattle by tonight. But instead we went through Dodd Narrows with the tide, to wind our way through the Canadian Gulf Islands on a lazy Indian summer afternoon. With the smell of burning leaves in the air, and a sloop reaching slowly down the sunny channel, those gray weeks up north seemed only a memory. With an hour or two of warm sun left, we tied to the float at Musgrave Landing at 4, and had a long walk in the leafy woods. Hiked to the ridge above, through open fields, and from there we could see the

*This tumble-down cannery on West Redonda Island is seen on an early November day.*

*Sam, the dog who is loved like a person, reposes in tall grass at Musgrave Landing.*

narrows laid out below us and Vancouver Island across the way. Sam ran and ran, back and forth across the fields in long loping strides, back to us, panting, his tongue hanging out, then off again. An old farm is up there on the ridge, and we filled our pockets with apples from the orchard.

The evening is still, the night starry; a fish jumped in the cove just now, and there was the far-away cry of a heron. Tomorrow we'll be in a town again, and the night after that, Seattle. Our season's over and our trip south just about done. In the spring it will start again. We're all tired now from the trip and the season, but still, it's sad to have it end. Our work is sometimes hard, and these trips tedious. At times the boat seems too small for the three of us. Still, life fills us up in a way it won't ashore. To take off in the black from some float or anchorage, and then run all day through this lonely country, to tie up early and walk the beach on a warm fall afternoon—we savor these times, and we motor a little slower as we get to the end—it happens every year.

*NOVEMBER 10*—The day came gray with a southeasterly breeze in the trees. I dawdled as long as I could, but finally cast off at 9. Into Boundary Passage, and across into the waters of Washington State. Summer homes on all the shores, but they're empty and boarded up for the winter. Stopped at Deer Harbor on Orcas Island. The floats were pulled up on the beach for the season, leaves swirling through the empty streets. A chill breeze swept the harbor; only a few boats are left on moorings. Ate lunch in an empty tavern with the hard afternoon light slanting down through the windows. Took off in a squall, water flying in the sheltered waters. Called a passing ferry for a check on the weather outside, and they said it was fine. I should have used my own judgment, I guess, but we were so close to the end that let my guard down.

We went through Lopez Pass and out into Rosario Strait, and in a minute things were doing, with wind against tide. Those seas must have been 12 feet high, just straight up and down. Gave the wheel to Susanna and went out to try and put the poles down and get the stabilizers in. It was pretty wild out there, and I was lucky to get the poles down without damage. Waves were coming from all directions—one came over the stern and very nearly took me with it. That was a dirty hour before we were across, and Susanna even wanted to turn around, and she's *never* said that before! It was just a rip, and the farther we went the easier it got. But we got caught with our pants down out there today, and we could have lost the whole business a few miles from the end. That southerly just never gives an inch—I thought I learned that lesson a long time ago, but I guess not. I didn't tell Susanna, but that was as close as we had come all season, and it was all over in just a few minutes.

Up Swinomish Slough at dusk, running at half speed and taking it all in. Just an hour before we had been fighting the tide rip; now we ran in the narrowest of channels, between grassy banks, with fields of tall grass flowing in the wind beyond. Dairy farms stretched away to the hills of the North Cascades in the distance, and raft after raft of ducks settled onto the still waters of the slough behind us in the rosy dusk. Here and there on the bank a few old houses stood with brightly lit windows; inside I could see people squinting out at us as we passed. I wondered what they thought on this windy, cold, November

*Above—A dairy farm on Swinomish Slough, Washington, north of Puget Sound. Right—At nearby La Conner is an old friend, the* W.B. Starr, *a vessel we have known for years.*

evening, when they looked out and saw these last stragglers still wandering down from Alaska.

Tied up at the float at La Conner, with just the faintest trace of color left in the sky to the west. This is a sleepy fishing town on a muddy slough; ahead and astern are tied silent seiners, their crews gone home, laid up for the season.

Uptown tonight for dinner, through empty streets, hardly seeing another person. Ate at a tavern over the water; two people at the bar and a solitary pool player at the tables. How strange it is to run through all that empty country up north, and now get here and find this all deserted, too. Walked back to the boat, our collars up against the wind, to find the frost thick on the float at 10.

*NOVEMBER 11*—Covered the last miles on a gray windy day. Off at 8 past the silent town, the channel twisting, the trees closing in. Then into Skagit Bay with a nasty short chop and muddy waters. Bucked south all day without another boat in sight until the last hour. The sound became narrower, and there were houses now on all the shores. Had a southeaster in our teeth right up to the breakwater and the locks beyond. The big steel gates of the locks closed behind the boat, and we went through alone in the smoky dusk, with the roar of Seattle traffic in the streets above. Sam went up to the bow to smell the new smells, and we tied to the dock at Fishermen's Terminal in Seattle, 20 days from Point Baker and almost seven months exactly from the day we left. How different from that bright, still morning when we passed out northbound, decks and quarters still piled high and the whole new season ahead of us.

*"When white men first started
fishing the Northwest Coast,
the rivers were choked with salmon,
and when the fish ran, the fishermen
worked seven days a week."*

# The Alaska Blues

It would be hard to be in the salmon fishery in Southeastern Alaska and not be aware of its constant decline. While billions are being spent to develop nonrenewable resources—minerals—Alaska's oldest renewable resource—salmon—is being poorly managed and underfunded. It is a bitter experience for the many persons who depend on the salmon fishery for a livelihood.

I first went to Southeastern Alaska in 1965, working as an engineer on a tender, buying fish. In my time off, I explored about half the bays in the region with our outboard. That lonely country left a deep impression on me, and I wanted to come back some day with my own boat.

In 1970 I leased a tired and rotting 30-foot gill-netter for the season in Alaska, but the engine blew before we made it to Canada.

I bought the *Doreen* in 1972—at last I was going north in my own boat. It was grand. We had a successful season, and on our way south to Seattle

*A beached gill-netter, pushed onto a bar in Chilkat Inlet by high winds.*

from fall fishing, Susanna and I stopped at Point Baker and found a piece of land and leased it from the state. The next spring, we loaded our boat with windows, doors and a hundred other things needed to build a cabin, and we ran up early and built a cabin before the fishing season started. There were fewer fish that year, and the fishing period for summer fishing was reduced to two days a week, but the price was high and we did well.

But 1974 was a disturbing year. We had our first one-day-a-week fishing periods for summer fishing. Then, in midseason, all net fishing in the region was closed down for 10 days because of the poor showing of fish. I was deeply worried about the future, so in the spring of 1975 I sold my boat and took a job as skipper on a big company-owned tender, buying fish again.

And 1975 turned out to be the worst year anyone had ever seen. The fishing was so poor in Sumner Strait and at Point Baker that they closed the area after the first week, and we moved south with our boats to Clarence Strait.

By July, several areas were closed to trollers, for the first time in anyone's memory, and at the end of the month, the entire region was closed to all net fishing for two weeks. I tied up the tender and flew south to lie on the beach until fishing opened again. That was a crazy feeling, taking a vacation in the middle of the season.

When the season reopened, we got lucky and scratched out a season in the last two months—Lynn Canal came through with a big run of fall dog salmon. But that's hard on your nerves, making your season in the last few weeks. Most people ended up paying their bills and having little left over. When the next season came, spring of 1976, I decided not to go north.

I don't know what the answers are. Some people say that all we need are a couple of good returns of fish up the creeks, and mild winters, and we'll be in the chips again. But I can't believe that. I know I've seen fewer and fewer fish every year, and I look at the statistics and see that's been the trend for more than 30 years. I've walked prime creeks at spawning time and seen hardly a fish. I've had my net set in the right spot at the right stage of the tide, with nobody between me and the ocean, and had hardly a fish. When I was doing poorly, I'd think the trouble was with me, that I was missing them. But now, after working on the tender and seeing what everybody else had, I know it's not just me. I think that the old management schemes simply aren't working any longer.

In Alaska, salmon management consists primarily of regulating the fishing time to ensure that enough fish enter the streams to provide a successful spawn for the next generation. The fish in the streams are estimated by aerial and shoreside counts, and the fishing periods are adjusted accordingly. But in 1975, some fishing areas were down to 12 hours a week, or were closed entirely, and still not enough fish returned to the streams to spawn.

*The tender* Emily Jane, *deck-loaded with dog salmon in Lynn Canal.*

*An abandoned herring-reduction plant at Port Walter near the southern tip of Baranof Island on Chatham Strait.*

When white men first started fishing the Northwest Coast, the rivers were choked with salmon, and when the fish ran, the fishermen worked seven days a week. The biggest problem was processing all the fish they caught. The runs dwindled, not only from the intensive fishing pressure but also from the effects of a growing population: Pollution from municipalities, industrial plants and poor logging practices fouled the spawning streams, and the dams built on rivers to generate electric power prevented the fish from getting upriver to spawn. The first step taken in fishery management was to reduce the fishing time.

But people in the coastal states and provinces soon realized that more direct steps needed to be taken. Therefore, hatcheries were built, and programs were established to rehabilitate the dwindling salmon runs and actually produce fish. Where spawning streams couldn't be restored, artificial spawning channels were built to provide better spawning conditions for the returning salmon. In many areas, especially in British

Columbia, these investments have paid off handsomely in increased catches.

Alaska has so far largely escaped the pressures of dam builders and pollution (the main exception being the pollution of streams caused by logging), and has looked forward to strong natural salmon runs. But the catch of salmon in Alaska has slowly declined. I believe, as do others, that fisheries management in Alaska can no longer be based so heavily on a program of simply limiting the catch. The time has come to make a bigger effort than ever before to rehabilitate the salmon fishery by means of hatcheries, artificial spawning channels or whatever technology seems indicated.

In one area, Prince William Sound, the local fishermen themselves have followed this path: they have funded and established their own hatchery program. This is one of the most hopeful signs for salmon fishing in all of Alaska. The governor of the state has stated his

*The old office at the Hood Bay cannery south of Angoon on Admiralty Island.*

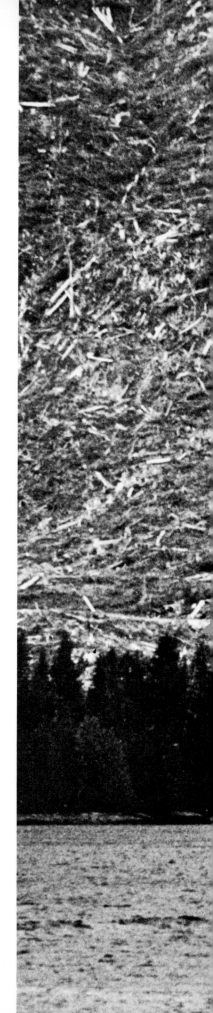

Kestrel *runs past cut-over slopes on El Capitan Passage.*

commitment to a fisheries rehabilitation program, funded by a tax on nonrenewable resources. We all hope he is able to follow through with it.

But perhaps even more serious than fisheries rehabilitation is the second challenge that fish and fishermen have to face in Southeastern Alaska: the conflict between logging and salmon production. Southeastern Alaska is almost entirely national forest, timber land owned and controlled by the federal government. In the 1950's, several large timber sales were made, with very good terms for the companies involved, to start a forest-products industry in the region and help move the state away from dependence on the seasonal fishing industry. Pulp mills and sawmills were built, and the first clear-cutting of trees ever permitted in the national forests was begun in the region. Today, the forest-products industry is one of the biggest employers in Alaska.

In many cases, because of the very favorable nature of the long-term contracts made with the timber companies in the fifties, the prices paid to the government for the timber are very low indeed. So, in a sense, the taxpayer is subsidizing the production of timber. In Southeastern Alaska, Japanese companies now have a large, if not controlling, interest in the timber industry, and virtually all of the timber cut is exported to Japan. So, in a sense, the taxpayer is subsidizing also the export of timber.

In 1973, the U.S. Forest Service announced its plans to start cutting timber around Point Baker and Port Protection. The two communities are on the northern tip of Prince of Wales Island, surrounded by hundreds of thousands of acres of virgin timber. The plan was to start cutting in 1975; a semipermanent logging camp would be built on the bay at Port Protection, and roads from the camp would go to both communities. Many residents of the area (I was one of them) were unhappy at the prospect, and we fought it tooth and nail. We formed the Point Baker Association and carried our fight ultimately to the office of the Secretary of Agriculture. Our protest was based on what we felt were violations of the Multiple Use-Sustained Yield Law (MU-SY) and the National Environmental Policy Act (NEPA). Most of our objections were disallowed; the logging began.

*The vessel* Hope, *beached at Point Baker, has been stripped piece by piece for firewood.*

The more we saw the effects of cutting practices on salmon production, the more we realized that the Forest Service was not being candid, and that real damage was being done to salmon streams.

In the spring of 1975, I flew over a recently logged valley at the head of Calder Bay, 20 miles south of Point Baker. I knew the valley before it was logged, and I had walked the creek when it was full of spawning salmon. What I saw from the airplane that day appalled me—and it flew in the face of all the assurances that the Forest Service had given us about careful logging. The valley had been logged from ridge to ridge, with hardly a tree left standing, and several downed trees were left in the bed of the creek. This kind of cutting meant increased stream temperatures, erosion and siltation, and heavier stream flows—all of which are harmful to spawning salmon and their fry.

We failed to get any satisfaction through administrative channels, so we took the Forest Service to court. We found, however, that neither of the new laws, MU-SY or NEPA, had teeth enough to do us any good. We would have lost the case had it not been for an old law, the Organic Act of 1897, which prohibited the cutting of other than mature trees in the national forests, thus prohibiting clear-cutting. Eventually Congress had to take up the question of reform, and one of our local fishermen went to Washington to testify before the appropriate committees. The committee members were duly shocked by his testimony, but the outcome of the reform bill is still far from certain. The big guns of the timber lobby are lined up against it.

It seems clear that logging and salmon production are going to have to exist in better harmony. We can't continue to fight, and we can't continue to develop one at the expense of the other.

When I first came to Southeastern Alaska in 1965, it was right on the heels of a year's fishing off northern Chile, the most barren, desolate and treeless coast that you could imagine. Standing on the deck of the tender making its way up the Inside Passage to Alaska, I thought, "This is it, God's country, right here—don't look any further." And then I started salmon fishing, and it was the best and the most enjoyable fishing that I'd ever done. But the fishing has changed for the worse faster than I would ever have thought it could, and I don't see any improvement soon.

The trend in the fishery is alarming. I feel that our way of life is threatened by forces over which we have little control.

So, as much as anything I want this book to be a record of the country we fished, our boats, and all of us.

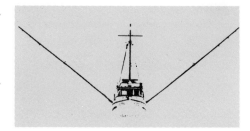

# Glossary:
## A Fisherman's Language

AWASH. Level with the sea or washed over by waves.

BACK. A wind shift counterclockwise, as from southerly to easterly. A wind shifting clockwise is said to *veer*.

BEACH. The shore; land.

BIGHT. An indentation in a shoreline, less indented than a bay or cove.

CHOP. Small, closely spaced waves produced by the action of wind on the water; a surface movement.

CORKLINE. The line at the top of a gill net to which the corks are fastened. See *gill net*.

DOWNWIND. In the direction toward which the wind is blowing; to leeward.

DRIFT. Debris in the water.

DROPPERS. Lines between corkline and weedline of a gill net. See *gill net*.

EBB. Change in depth of water from high point to low point, with the flow or current passing out to sea, draining the bays. See also *tidal current*.

ESCAPEMENT. Portion of a run of spawning fish that is allowed to pass through a fishing area, and escape being caught, so that they can enter streams where they will spawn.

FATHOM. A measurement of depth: 6 feet.

FLOOD. Change in depth of water from low point to high point, with the flow or current passing inshore, filling bays. See also *tidal current*.

FO'C'SLE. Forecastle. The space in the bow of a fishing boat, under the foredeck, where the crew's living quarters are.

GILL NET. A net designed to catch fish by their gills; the mesh is nearly invisible, and the mesh size is large enough to allow the head, but not the after body, of the fish to pass through. The gill net hangs in the water vertically, like a fence. The net consists of

the *web,* or *net,* a *corkline* at the top and a *leadline* at the
bottom; a net made for use in Southeastern Alaska also has a
*weedline* at the top of the web. The corkline is the line, usually
7/16-inch-diameter polypropylene, to which the corks that
provide flotation for the net are fastened. From the corkline,
much lighter lines called *droppers* (1/4-inch-diameter or less, 6 to
12 inches long, spaced every 3 to 6 feet) extend to the weedline at
the top of the web; suspending the net in this manner leaves a
largely open area at the top, through which the debris found in
Alaskan waters can pass without fouling the net. At the bottom
of the net is the leadline, 3/8-inch diameter, to which the lead
weights are fastened. Almost all modern nets, however, use lead-
core line instead of line and separate lead weights. The web of the
net is woven (not monofilament) nylon in colors designed to
make the net as invisible as possible in various circumstances.
The mesh sizes used for salmon in Southeastern Alaska generally
range from 5 inches to 6-7/8 inches; this is the maximum
diagonal measurement from corner to opposite corner, with the
mesh pulled tight between corners. Gill nets may be streamed
from the boat, allowed to drift free of the boat, anchored, or set
with one end fast to shore; they may be fished either near the top
surface of the water or on the bottom. In Southeastern Alaska,
salmon gill-netting is done almost exclusively with nets fished on
the surface and allowed to drift; anchoring of the net is
prohibited.

GROUND SWELL. See *swell.*

GURDY. Small winch, usually power-driven, for raising and
lowering a trolling line.

GUT. A narrow passage or channel.

HAND-JIGGING GEAR. Hook-and-line fishing tackle, consisting
chiefly of artificial lures fished close to the bottom and moved up

and down in a manner designed to attract the attention of fish.
The method is more appropriate to an individual fishing for
subsistence than a commercial enterprise.

HANG A NET. Construct a net by sewing the web to the corkline and
leadline.

HOOK. Anchor. To *drop the hook* is to anchor. To *pull the hook* is
to lift anchor.

HOUSE. Wheelhouse or pilothouse of a vessel.

HULL-DOWN. A vessel on the horizon, with the hull out of sight
below the horizon and only the superstructure visible.

INSIDE. In sheltered waters of the Inside Passage, rather than on
the open ocean.

JOGGING. Operating a vessel with the lowest possible engine speed,
short of losing steerage way—a vessel's equivalent to a
person's treading water. Usually done only in severe weather.

JUNKED OUT. Forced to haul in a gill net because it is clogged with
debris.

KEEP WAY ON. Keep moving. Usually, but not necessarily, refers to
forward motion.

KNOT. A unit of speed. One knot is one nautical mile, or 1.151
statute miles, per hour; 10 knots is 11.51 miles per hour. This
is so because the nautical mile (6,076 feet) is longer than the
statute mile (5,280 feet) used on land.

LEADER. A short length of wire or line by which a hook is attached
to a fishing line.

LEADLINE. A lead-weighted line fastened to the bottom of a net.
The leadline may be made up of individual lead weights, or
lead-core line may be used. See *gill net.*

LEE. In a general sense, the side away from, or sheltered from, the
wind. On a vessel, the side toward the wind is the *windward
("wind'ard")* side; the side away from the wind is the *leeward*

234

(*"loo'rd"*) side. Anything—a small boat perhaps—so close by the leeward side of the vessel as to be in its wind shadow is said to be *in the lee* of the vessel. In the same way, a vessel on the sheltered side of an island is *in the lee* of the island. But when a vessel is blown toward shore by the wind, the shore—certainly unsheltered—is a *lee shore.*

LONGLINE. A long length of line to which is fastened a series of hooks on thin, wire leaders. In Alaska, longline gear is used for bottom fishing, primarily for halibut; in other areas, longline gear may be fished on the surface for tuna, swordfish or shark.

MINUS TIDE. An unusually low tide.

MUG-UP. A cup of coffee and a snack or sandwich.

OUTSIDE. On waters of the open ocean, rather than within the protection of the Inside Passage.

OVERFALL. A confused sea caused by the collision of strong tidal currents in a narrow channel. The resulting overfall is seen as a four- to five-foot drop in the height of the surface of the water over a short distance, as if the water were dropping over a shelf on the bottom. The phenomenon bears some relation to a tide rip.

PACKER. A boat that buys fish from the fishermen on the fishing grounds.

PICKING. Removing fish from a gill net as the net is hauled into the boat.

POWER BLOCK. A pulleylike device, usually operated by hydraulic power, that pulls a purse seine out of the water and lowers it onto the deck.

PRAM. A small boat, usually shorter than 12 feet, with a blunt bow instead of a pointed bow.

PURSE SEINE. A net used to encircle a school of fish. The bottom of the net is drawn closed by a purse line, making the net take the

shape of a basket, before the net is hauled to the boat with its
catch.

QUARTER (NOUN). The aftermost part of the side of a vessel, a rear
corner, left or right of the stern.

QUARTER (VERB). To go diagonally into a wind or sea, so as to take
its force neither head-on nor at right angles.

QUEER ONE. A big maverick wave that catches the boatman
unaware.

RANGE. A pair of fixed objects seen in line with one another, with
the line of sight between them expressed as a true compass
direction. To *run ranges* is to pilot a vessel in accordance with the
information gotten from ranges.

RIP. Short, steep waves caused by the meeting of conflicting
currents.

RUN. A group of fish on their way to spawn.

RUN THE NET. Operate the boat along the length of the net in order
to inspect the net and see whether it has fish in it.

RUNNING. A vessel's being under way.

SALMON NAMES. Salmon called by one name in Alaska may be called
by other names elsewhere:

King   Chinook, tyee or spring
Silver   Coho or silversides
Red   Sockeye or blueback
Pink   Humpback or humpie
Chum   Dog

SCOPE. Ratio between length of anchor line and depth of water.
Ratios between 4 and 7 to 1 are considered desirable, so that a
vessel would pay out 40 to 70 feet of line to anchor securely in 10
feet of water.

SEINER. A fishing boat equipped with a purse seine net.

SET. The deployment of a net.

SILVERS. Silver salmon. See *salmon names.*

SKATE, SKATE OF GEAR. A unit of a longline; usually 300 fathoms
   long and having from75 to nearly 200 hooks.
SKUNK DAY. A day when no fish have been caught.
SLOUGH. A waterway through a tidal marsh.
SOUNDER. Depth sounder, an electronic device for measuring the
   depth of water.
SPRING LINES. Mooring lines so arranged as to permit the vessel to
   float freely in any direction except ahead or astern. Spring lines
   lead from the bow aft and from the stern forward; they are
   useful in docking and undocking because with them the vessel
   can be made to move sideways in or out of a limited space.
STABILIZERS. Also called flopper-stoppers and, by the technically
   minded, paravanes. These stabilizers are delta-wing foils, each
   having a weight on the nose and a fin on top, that are trailed
   from the trolling poles to help keep the boat from rolling.
SWELL. A long, heavy undulation of the sea, resulting in a deep,
   rolling wave.
TENDER. A packer (see).
TIDAL CURRENT. Flow of water from place to place caused by the
   rise and fall of the tide. See also *ebb* and *flood*.
TOW. The burden (usually a loaded barge) pulled by a tow boat
   (tug). A double tow is two barges pulled by one tow boat.
TROLL. To fish by trailing a hook and line; the hook may be baited
   with fresh bait or be part of an artificial lure.
TURN, THE TURN. The change of tide.
WEEDLINE. A 1/4-inch polypropylene line fastened along the top of
   the mesh of a gill net. Use of a weedline allows the net to be
   suspended a few feet below the much heavier corkline (7/16-inch
   diameter) in order to let floating debris pass through the top of
   the net without fouling it. This construction is common in
   Alaska. See also *gill net.*